ELECTRICAL ENGINEERING DEVELOPMENTS

LITHIUM BATTERIES: RESEARCH, TECHNOLOGY AND APPLICATIONS

ELECTRICAL ENGINEERING DEVELOPMENTS

Additional books in this series can be found on Nova's website at:

https://www.novapublishers.com/catalog/product_info.php?cPath=23_29&products_id=10199

Additional -books in this series can be found on Nova's website at:

https://www.novapublishers.com/catalog/index.php?cPath=23_29&seriespe=Electrical+Engineering+Developments

ELECTRICAL ENGINEERING DEVELOPMENTS

LITHIUM BATTERIES: RESEARCH, TECHNOLOGY AND APPLICATIONS

GREGER R. DAHLIN
AND
KALLE E. STROM
EDITORS

Nova Science Publishers, Inc.
New York

For permission to use material from this book please contact us:
Telephone 631-231-7269; Fax 631-231-8175
Web Site: http://www.novapublishers.com

LIBRARY OF CONGRESS CATALOGING-IN-PUBLICATION DATA

Lithium batteries : research, technology, and applications / editors, Greger
R. Dahlin and Kalle E. Strom.
p. cm.
Includes index.
ISBN 978-1-60741-722-4 (hardcover)
1. Lithium cells. I. Dahlin, Greger R. II. Strom, Kalle E.
TK2945.L58L5535 2009
621.31'2423--dc22

2009051579

Published by Nova Science Publishers, Inc. ✢ New York

CONTENTS

PREFACE

Lithium ion batteries, a class of chemical power sources that use an electrochemical process of lithium ion intercalation into or de-intercalation from host materials, are gaining dominance in mobile electronic applications, and are also showing promise for an upcoming new generation of electric vehicle applications. Since SONY Corporation commercialized rechargeable lithium-ion batteries, the batteries have been widely utilized as the power sources in a wide range of applications, such as mobile phones, laptop computers, digital cameras, electrical vehicles, and hybrid electrical vehicles. This book is concerned with the recent developments in and research of LiFePo4 cathode materials with an emphasis on the synthesis method and how to improve electrochemical performance. Moreover, the efforts made to develop other new inorganic cathode materials for a new generation of lithium ion batteries are reviewed. A systematic semi-empirical way to analyze the constituents of total cell impedance in lithium-ion battery is also presented. In addition, overcharge protection is not only critical for preventing the thermal runaway of lithium-ion batteries during operation, but also important for automatic capacity during battery manufacturing and repair. This book compares three overcharge protection strategies - external circuit protection, inactivation agents, and redox shuttles - to highlight the advantage of redox shuttles for overcharge protection. The safety of lithium-ion battery packs are also discussed, as well as techniques for studying thermal stability, such as differential scanning calorimetry and accelerating rate calorimetry.

Chapter 1-Since SONY Corporation commercialized rechargeable lithium-ion batteries 18 years ago [1], the batteries have been widely utilized as the power sources in a wide range of applications, such as mobile phones, laptop computers, digital cameras, electrical vehicles, and hybrid electrical vehicles. In rechargeable lithium-ion batteries, cathode materials are one of the key components, and mainly devoted to the performance of the batteries. Among the known cathode materials, layered $LiCoO_2$, $LiMnO_2$, and $LiNiO_2$, spinel $LiMn_2O_4$, and other cathode materials such as elemental sulfur have been studied extensively [2-15] while $LiCoO_2$ has been used as the cathode material for commercial lithium-ion batteries. However, due to the toxicity and the high cost of Co, novel cathode materials must be developed not only in relation to battery performance but also in relation to safety and cost.

Chapter 2- Lithium ion batteries, a class of chemical power sources that use an electrochemical process of lithium ion intercalation into or de-intercalation from host materials, are gaining dominance in mobile electronic applications, and also showing promise for an upcoming new generation of electric vehicle applications. Currently, the most successful active electrode materials used in lithium ion batteries are graphite (anode material with a specific capacity of 350 mA h g^{-1}) and $LiCoO_2$ (cathode material with a specific capacity of 135 mA h g^{-1}). Under the driving force of safety issues, a new cathode material, $LiFePO_4$, has been developed in recent years as the most promising cathode material for next-generation lithium ion batteries. However, this new material can deliver a specific capacity of only 150 mA h g^{-1}, which is far less than that of anode materials (Figure 1) [[1]]. The low specific capacity of cathode materials has been identified as the factor preventing lithium ion batteries from meeting the high capacity and high power demands of automobiles and electronic devices. Therefore, finding cathode materials with higher specific capacities has become the key priority in lithium ion battery research and development (R&D).

Chapter 3- This work presents a systematic semi-empirical way to analyze the constituents of total cell impedance in a lithium-ion battery, and their time-dependent contributions to total direct current (dc) polarization. The approach includes the differentiation of internal resistive elements, followed by theoretical calculations of their contributions to total polarization using circuit analysis. Our method provides a fast and reliable way to design a high-power battery with the instantaneous input/output power that best fits the user's specific needs. It also provides insight into the design of high-power with long shelf life and calendar life. We begin with an overview of high-power cell design. Methodology to differentiate and quantify the time-dependent contribution of elementary resistances to total polarization is given, and applications to power aging in battery use, and power decline at low operating temperature, are suggested. A strategy for the design of materials to meet power requirements is discussed for each case.

Chapter 4- Overcharge protection is not only critical for preventing the thermal runaway of lithium-ion batteries, but also important for automatic capacity balancing. This chapter compares three overcharge protection strategies—external circuit protection, inactivation agents, and redox shuttles—to highlight the advantage of redox shuttles for overcharge protection. Then the redox shuttle history and mechanism are introduced and the latest advances on redox shuttles are described. Fundamental studies for designing stable redox shuttles for use in lithium-ion batteries are also discussed.

Chapter 5- Parallel to the rising market for lithium-ion power packs, more incidents of severely debilitating and sometimes fatal tragedies, as a result of battery hazards are being reported. Some of the safety risks of lithium-ion batteries are inherent in the fact that they combine highly energetic materials that are in contact with a flammable electrolyte based on organic solvents. Moreover, the potential ranges experienced by these cells are beyond the thermodynamic stability windows of the electrolytes, which can decompose upon contact with the charged active materials. The interface between the cathode and electrolyte is of special concern since partial dissolution of the active material can create further complications. This chapter discusses processes at the positive electrode that can lead to thermal runaway, especially at those based on the most popular cathode materials, $LiCoO_2$ and $LiNi_{0.8}Co_{0.2}O_2$. Measures such as coating cathode particles with inert oxides have been shown to improve cell safety by increasing the onset temperature of electrode-electrolyte reactions and lowering the exothermicity of such reactions. Additionally, coatings also bestow improved cyclability to the cathodes. Reactivity of cathode active materials is also related to electrolyte composition. Electrolyte additives and non-flammable electrolytes are a case in point. Techniques for studying thermal stability such as differential scanning calorimetry and accelerating rate calorimetry are also discussed.

Chapter 6- As is well known [1-3], the lithium battery is a rechargeable battery and its lithium comes from the cathode electrode materials such as lithium intercalated transition metal oxides, for example, $LiCoO_2$, $LiNiO_2$, $LiMnO_2$, $Li(Co_{1-x-y}Ni_xMn_y)O_2$, and $LiMn_2O_4$. During the charging process, the Li^+ ions pull out from the lithium intercalated oxides by electric field and are expelled into the carbon layers of graphite anode through an electrolyte. Therefore, the graphite anode acts as Li^+ ions sink. However, during the discharging process, the Li^+ ions stored in the graphite anode act as Li+ source and will flow out from the graphite anode intercalating into the oxygen closed packed layers of the dioxide cathode through an electrolyte. So, the cathode is a sink of the flowing Li^+ ions. The anode and cathode both act as Li^+ ion source and sink. The capacity of the lithium battery is dominated by the Li^+ ion source storage capacity and the sink volume. The rate of Li^+ ions flow is also related to the source and sink capability. The cathode and anode, especially the cathode, are very important for the lithium battery.

Chapter 7- Lithium vanadium oxides have been extensively studied because of their possible application as a cathode material for rechargeable lithium batteries. Due to their low cost, they are one of the promising substitutes for the expensive $LiCoO_2$ cathode presently commercially used. Lithium vanadium oxides including γ-LiV_2O_5 and LiV_3O_8 have been prepared by soft solution methods in this study. In the first part of this work, γ-LiV_2O_5 nanorods have been prepared directly by a simple solvothermal method using ethanol as a solvent, which also serves as a reducing agent. The γ-LiV_2O_5 nanorods with diameters of 30-40 nm obtained at 160 °C shows a larger capacity of 259 mAh/g in the range of 1.5 - 4.2 V, and its capacity remained 199 mAh/g after 20 cycles. In the second part, LiV_3O_8 nanorods have been obtained by a novel hydrothermal-based two-step method. The LiV_3O_8 sample treated at 300 °C shows a poor crystallinity while a specific capacity of 302 mAh/g in the range of 1.8 - 4.0 V, and its capacity remained 278 mAh/g after 30 cycles. It indicates that the lithium vanadium oxide nanorods prepared by the above methods have potentiality to be used as cathode material in rechargeable lithium ion batteries.

Chapter 8- In this study, electric-drive vehicles with series powertrains were configured to utilize a lithium- ion battery of very high power and achieve sport-sedan performance and excellent fuel economy. The battery electrode materials are LiMn2O4 and Li4Ti5O12, which provide a cell area-specific impedance of about 40% of that of the commonly available lithium-ion batteries. Data provided by EnerDel Corp. for this system demonstrate this low impedance and also a long cycle life at 55°C. The batteries for these vehicles were designed to deliver 100 kW of power at 90% open- circuit voltage to provide high battery efficiency (97-98%) during vehicle operation. This results in battery heating of only 1.6°C per hour of travel on the urban dynamometer driving schedule (UDDS) cycle, which essentially eliminates the need for battery cooling. Three vehicles were designed, each with series powertrains and simulation test weights between 1575 and 1633 kg: a hybrid electric vehicle (HEV) with a 45-kg battery, a plug-in HEV with a 10-mile electric range (PHEV10) with a 60-kg battery, and a PHEV20 with a 100-kg battery. Vehicle simulation tests on the Argonne National Laboratory's simulation software, the Powertrain System Analysis Toolkit (PSAT), which was developed with MATLAB/Simulink, showed that these vehicles could accelerate to 60 mph in 6.2 to 6.3 seconds and achieve fuel economies of 50 to 54 mpg on the UDDS and highway fuel economy test (HWFET) cycles. This type of vehicle shows promise of having a moderate cost if it is mass produced, because there is no transmission, the engine and generator may be less expensive since they are designed to operate at only one speed, and the battery electrode materials are inexpensive.

In: Lithium Batteries: Research, Technology... ISBN: 978-1-60741-722-4
Editors: G. R. Dahlin and K. E. Strom

Chapter 1

$LiFePO_4$ CATHODE MATERIALS FOR LITHIUM-ION BATTERIES

B. Jin** and Q. Jiang

Key Laboratory of Automobile Materials (Jilin University),
Ministry of Education, and School of Materials Science and Engineering,
Jilin University, Changchun 130025, China.

1. INTRODUCTION

Since SONY Corporation commercialized rechargeable lithium-ion batteries 18 years ago [1], the batteries have been widely utilized as the power sources in a wide range of applications, such as mobile phones, laptop computers, digital cameras, electrical vehicles, and hybrid electrical vehicles. In rechargeable lithium-ion batteries, cathode materials are one of the key components, and mainly devoted to the performance of the batteries. Among the known cathode materials, layered $LiCoO_2$, $LiMnO_2$, and $LiNiO_2$, spinel $LiMn_2O_4$, and other cathode materials such as elemental sulfur have been studied extensively [2-15] while $LiCoO_2$ has been used as the cathode material for commercial lithium-ion batteries. However, due to the toxicity and the high cost of Co, novel cathode materials must be developed not only in relation to battery performance but also in relation to safety and cost.

* Corresponding author: Tel.: +86-431-85095170; E-mail: jinbo@jlu.edu.cn (B. Jin)

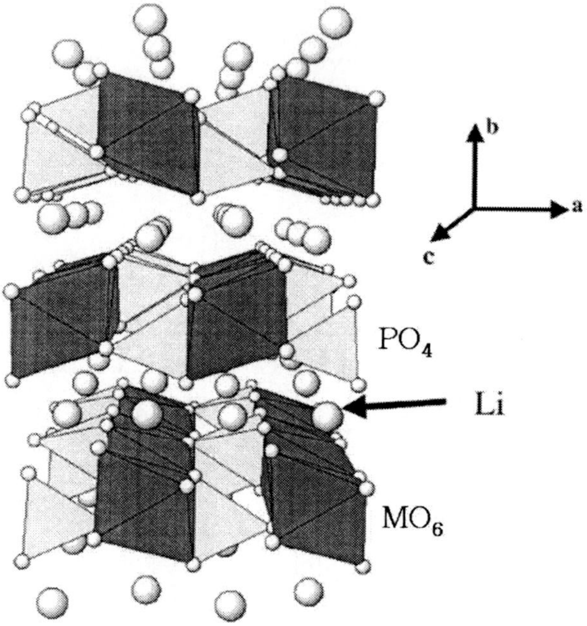

Figure 1. The schematic representation of the crystal structure of $LiMPO_4$ (M=Fe, Mn, Co, and Ni) compounds showing the HCP oxygen array with MO_6 and PO_4 groups.

Recently, $LiMPO_4$ (M = Fe, Mn, Ni, and Co) proposed by Goodenough *et al.* with an ordered olivine-type structure has attracted an extensive attention due to a high theoretical specific capacity (~170 mAh g^{-1}) [16-35]. As shown in Figure 1, $LiMPO_4$ (M = Fe, Mn, Co, and Ni) adopts an olivine-related structure, which consists of a hexagonal closed-packing (HCP) of oxygen atoms with Li^+ and M^{2+} cations located in half of the octahedral sites and P^{5+} cations in 1/8 of tetrahedral sites. This structure may be described as chains (along the *c* direction) of edge-sharing MO_6 octahedra that are cross-linked by the PO_4 groups forming a three-dimensional network. Tunnels perpendicular to the [010] and [001] directions contain octahedrally coordinated Li^+ cations (along the *b* axis), which are mobile in these cavities. Among these phosphates, $LiFePO_4$ is the most attractive because of its high stability, low cost and high compatibility with environments [36-37]. However, it is difficult to attain the full capacity because the electronic conductivity is very low, which leads to initial capacity loss and poor rate capability, and diffusion of Li^+ ion across the $LiFePO_4$/$FePO_4$ boundary is slow due to its intrinsic character [16]. The electronic conductivity of $LiFePO_4$ is only 10^{-9}-10^{-10} S cm$^-$

[1] [38], being much lower than those of $LiCoO_2$ (~10^{-3} S cm^{-1}) and $LiMn_2O_4$ (2×10^{-5}-5×10^{-5} S cm^{-1}) [39-40]. Many researchers have suggested solutions to this problem as follows: (i) coating with a conductive layer around the particles [41-42]; (ii) ionic substitution to enhance the electrochemical properties [38]; and (iii) synthesis of particles with well-defined morphology [43-44]. The most researches focus on synthesis method and developing the simple preparation procedure to improve low electronic conductivity and cycling performance of $LiFePO_4$.

This review will be concerned with the recent development and research of $LiFePO_4$ cathode materials with emphasis on synthesis method and how to improve electrochemical performance. Here we will also draw the cathode performance from examples taken from our own work. This contribution consists of five sections. Section 1 is entitled Introduction. The following section (Section 2) describes the synthesis method. Section 3 focuses on how to improve electrochemical performance. Section 4 provides summary and future prospects. Section 5 is acknowledgments.

2. Synthesis Method of LiFePO4 Cathode Materials

2.1. Solid-State Reaction

Many research groups have tried to use solid-state reactions to synthesize $LiFePO_4$ [16, 45-49]. The solid-state reaction is a conventional synthesis method, which usually needs a two-step heating treatment including the first firing in a temperature range of 300-400 °C and subsequent one between 600 and 800 °C. These repeated heat-treatments result in a large particle size due to crystal growths in the final product [43, 45]. Goodenough *et al.* [16] synthesize $LiFePO_4$ by direct solid-state reaction of stoichiometric amounts of Fe(II)-acetates, ammonium phosphate, and Li carbonate. The intimately ground stoichiometric mixture of the starting materials is first decomposed at 300 to 350 °C to drive away the gases. The mixture is then reground and returned to the furnace at 800 °C for 24 h before being cooled slowly to room temperature. The X-ray diffraction (XRD) testing shows the emergence and growth of a second phase at the expense of $LiFePO_4$ synthesized by the above solid-state reaction as more and more Li ions are extracted. With total chemical delithiation, the second phase could be identified by both chemical analysis and Rietveld refinement to XRD data to be $FePO_4$. The XRD testing for chemical lithiation of $FePO_4$ shows the emergence and growth of $LiFePO_4$ at the expense of $FePO_4$ on more lithiation. Both $LiFePO_4$ and $FePO_4$ have the same space group. There are contractions of *a* and *b* constants on chemical extraction of Li from $LiFePO_4$, but a small increase in *c* constant. The volume decreases by 6.81% and the density increases by 2.59%. Electrochemical charge and discharge testing results indicate that approximately 0.6 Li atoms per formula unit can be extracted at a closed-circuit voltage of 3.5 V vs. Li and the same amount can be reversibly inserted back into the structure on discharge. The extraction and insertion of Li ions into the structure of $LiFePO_4$ is not only reversible on repeated cycling; the capacity actually increases slightly with cycling.

Kim *et al.* [49] synthesize Li_xFePO_4 (X = 0.7-1.1) by a solid-state reaction. Li_2CO_3, $FeC_2O_4{\cdot}2H_2O$ and $NH_4{\cdot}H_2PO_4$ as starting materials are milled with ZrO_2 ball in acetone for 24 h. After acetone is removed, the mixture is then decomposed at 350 °C for 10 h in flowing N_2 gas to avoid oxidation of Fe^{2+}. The powder is ground again using mortar and pestle, then it is pelletized. Finally the samples are heated at 700 °C for 24 h in flowing N_2 gas. The lattice parameters calculations of Li_xFePO_4 synthesized via the above solid-state reaction process with different Li contents demonstrate that lattice constants of these samples are approximately similar. Comparison of discharge capacities of Li_XFePO_4 with various current densities presents that $Li_{0.9}FePO_4$ has more capacity and better rate capability than the other two samples.

2.2. Hydrothermal Method

The hydrothermal synthesis is a useful method to prepare fine particles, and has some advantages such as simple synthesis process, and low energy consumption, compared to high firing temperature and long firing time during solid-state reaction used conventionally [50-56]. We also report the synthesis of $LiFePO_4$ by the hydrothermal synthesis [57-60]. Although $LiFePO_4$ can be easily synthesized hydrothermally at 150-220 °C and its XRD pattern looks good, it gives poor cycling performance; The HR-TEM image of $LiFePO_4$ heat-treated at 170 °C and subsequent at 500 °C in Figure 2 displays that amorphous layers with a thickness of about 1-3 nm are coated on the particle surfaces due to generation of carbon on the particle surfaces through decomposition of ascorbic acid as a reducing agent during the hydrothermal reaction, which results in an increase in the discharge capacity as demonstrated in Figure 3.

Whittingham *et al.* [52] also demonstrate hydrothermal synthesis of $LiFePO_4$ where the used starting materials are $FeSO_4 \cdot 7H_2O$, H_3PO_4 and LiOH. The molar ratio of the Li:Fe:P is 3:1:1, and a typical concentration of $FeSO_4$ is 22 g/liter of water. Sugar and/or L-ascorbic acid are added as an in situ reducing agent to minimize the oxidation of ferrous to ferric. Multi-wall carbon nanotubes are also added to improve electronic conductivity of $LiFePO_4$. The resulting grayish blue gel is transferred into a 125 ml capacity Teflon-lined stainless steel autoclave, which is sealed and heated at 150-220 °C for 5 h. Precipitates are collected by suction filtration and dried at 60 °C for 3 h in the vacuum oven. The XRD results demonstrate that the only phase observed is $LiFePO_4$. The lattice constants obtained from Rietveld refinement are: a = 10.332(2) Å, b = 6.005(1) Å, c = 4.6939(6) Å, V = 291.2 $Å^3$. Charge/discharge tests results in the first cycle show that for $LiFePO_4$ synthesized by the above hydrothermal synthesis, close to 160 mAh g^{-1} capacity is obtained on the charging cycle, and the capacity is over 145 mAh g^{-1} on discharge which is maintained over subsequent cycling.

2.3. Co-Precipitation

The co-precipitation procedure, a commercially feasible process, can prepare a fine, chemically uniform and more homogenous powder size distribution of $LiFePO_4$. Yang *et al.* [61] prepare $LiFePO_4$ with co-precipitation from aqueous solution containing trivalent iron ion. The aqueous precursor mixture of $Fe(NO_3)_3$, $LiNO_3$, $(NH_4)_2HPO_4$, ascorbic acid and appropriate amount of ammonia is used. The purpose of ascorbic acid has reduced Fe^{3+} to Fe^{2+} in the aqueous precursor. The amount of sugar added into the precursor solution is 20 wt % of $LiFePO_4$ to be formed. The co-precipitated powder can be easily separated in a centrifuge and then the co-precipitated powder is dispersed in the hydrolyzed sugar solution, followed by drying and heating. The sugar-coated powder is calcined at 350 °C for 10 h and subsequently sintered at 600 °C for 16 h in N_2 atmosphere. The sugar will be converted to carbon and distributed evenly on the $LiFePO_4$ powders. The particle size distribution result of $LiFePO_4$ synthesized via the above co-precipitation process shows that the particle distribution is bimodal, the population peak around smaller particle size is $LiFePO_4$ powder (about 1.51 μm) and another population peak at larger particle size (about 8.04 μm) can be attributed to the $LiFePO_4/C$ particles composed porous carbon structure with $LiFePO_4$ embedded. The charge/discharge test results demonstrate that $LiFePO_4/C$ synthesized via the above co-precipitation process can exhibit good capacity retention with slow charge/discharge rate (C/10-C/3), 85% of theory capacity of 169 mAh g^{-1}.

Park [62], Arnold [63], Ni [64], Park [65] and Prosini [66] *et al.* also improve the electrochemical performance of $LiFePO_4$ by co-precipitation method.

2.4. Emulsion-Drying Method

$LiFePO_4$ can be prepared via a hydrothermal method as mentioned above, but encounters the problem that some Fe ions reside on the Li sites and therefore deteriorates cell properties [67]. In such a liquid-phase synthesis, a solid phase is usually formed through a chemical reaction in the liquid phase. Hence, compared with solid-state reaction methods, some advantages are expected for the resultant powders, such as homogeneous mixing, lower heating temperature and smaller particle sizes. Emulsion-drying method as a new liquid-phase synthesis route is also used to prepare olivine-type $LiFePO_4$. Myung *et al.* [37] prepare $LiFePO_4/C$ composite by emulsion-drying method. Stoichiometric amounts of $LiNO_3$, $Fe(NO_3)_3{\cdot}9H_2O$ and $(NH_4)_2HPO_4$ are dissolved in distilled water. The aqueous solution is then vigorously mixed with a mixture of an oily phase, Kerosene : Tween 85 (surfactant) = 7 : 3 in volume, to prepare a homogeneous water-in-oil (W/O) type emulsion, in which cations are distributed very uniformly on an atomic scale. Finally, the prepared W/O type emulsion consisting of $LiNO_3$, $Fe(NO_3)_3{\cdot}9H_2O$, and $(NH_4)_2HPO_4$ is mainly composed of an oil phase (aqueous : oil phases = 2 : 8 in volume). The emulsion-dried precursor is burned out at 300 or 400 °C with a certain time in an air-limited box furnace. The obtained powders are then calcined at the desired temperatures for a specific time in a tube furnace with an Ar atmosphere. The charge/discharge testing results of $LiFePO_4/C$ composite synthesized via the above emulsion-drying process and cycled at 50 °C indicate that a higher capacity of about 140 mAh g^{-1} is obviously observed at 50 °C and the capacity retention during cycling is over 98%.

Chung *et al.* synthesize $LiFePO_4$ by direct heating of a dried emulsion precursor [68]. $LiNO_3$, $Fe(NO_3)_3{\cdot}9H_2O$ and $(NH_4)_2HPO_4$ are used as the starting materials. The dried emulsion precursor powders are heated under Ar flow at a heating rate of 5 °C/min to different temperatures. The cycle performance of $LiFePO_4$ synthesized at various temperatures and at 750 °C with 40 wt % carbon black as a conductive agent via the above emulsion-drying process demonstrate that the capacity obtained from the compound heated at 750 °C is higher than that obtained at 850 °C due to the particle-size effect, and the initial discharge capacity of $LiFePO_4$ synthesized at 750 °C with 40 wt % carbon black is 132.5 mAh g^{-1}, and increases to 151 mAh g^{-1} at the 10th cycle due to an enhancement in electronic conductivity through the use of a large amount of carbon black.

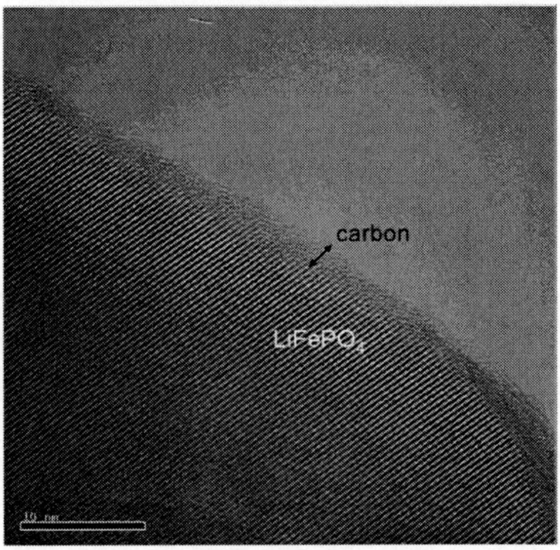

Figure 2. The HR-TEM image of $LiFePO_4$ heat-treated at 170 °C and subsequent 500 °C.

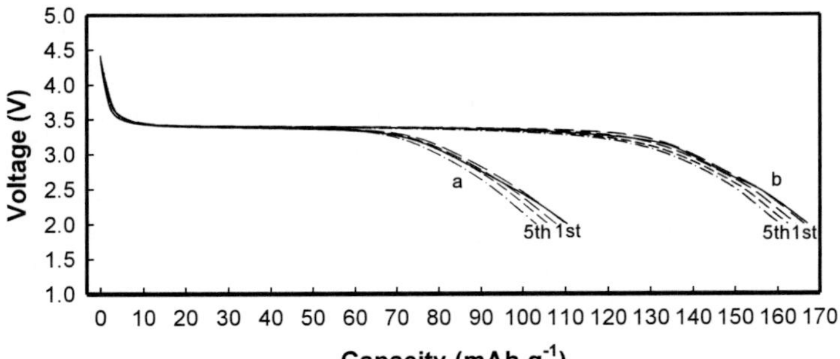

Figure 3. The discharge curves of $LiFePO_4$ synthesized at (a) 170 °C and (b) 170 °C and subsequent 500 °C.

2.5. Sol-Gel Method

There has been much interest recently in $LiFePO_4$ made by a sol-gel process [69-76]. Gaberscek *et al.* [73] synthesize $LiFePO_4$-based composite materials via a sol-gel method. 0.01 mol of Li_3PO_4 and 0.02 mol of H_3PO_4 are dissolved in 200 mL water by stirring at 70 °C for 1 h separately. 0.03 mol of iron (III) citrate is dissolved in 300 mL of water by stirring at 60 °C for 1 h. The two solutions are mixed together and dried at 60 °C for 24 h. After thorough grinding with a mortar and pestle, the obtained material is fired in inert (Ar) or reductive (5 % of H_2 in Ar) atmosphere at 500-700 °C for 15 min-72 h. The resulting $LiFePO_4/C$ consists of micrometer-sized particles containing pores with wide distribution of sizes. When filled with electrolyte, the pores are responsible for supply of ions while the distance between the pores (30-150 nm) determines the solid-state diffusion kinetics. The walls of pores are covered with a carbon layer, which serves as an electron conductor and is thin enough (2-3 nm) to allow penetration of Li ions. The electrochemical test data demonstrate that $LiFePO_4/C$ synthesized via the above sol-gel process at lower rates can recover towards the nominal capacity even after 50 cycles of the very high rate operation of 3400 mA/g.

Choi *et al.* [71] also report the synthesis of olivine-type $LiFePO_4$ by a sol-gel route using lauric acid as the surfactant while $CH_3CO_2Li{\cdot}2H_2O$, $FeCl_2{\cdot}4H_2O$ and P_2O_5 are used as the starting materials. Each precursor is dissolved separately in ethanol to yield a 1 M solution. Fe and P solutions are first mixed in the desired stoichiometric ratio and stirred for 3 h followed by the addition of stoichiometric amount of the Li solution. Equal molar ratio of lauric acid surfactant is added to the solution after 3 h of stirring. After 4 h, the reaction is presumed to be complete and the ethanol is evaporated under continuous flow of ultra high purity-Ar followed by heat-treatment under H_2/Ar = 10%/90% atmosphere at 500 °C for 5 h to prevent the possible formation of Fe^{3+} impurities. $LiFePO_4$ synthesized with lauric acid surfactant via the above sol-gel process can deliver a specific capacity of 125 and 157 mAh g^{-1} at discharge rates of 10 and 1C with less than 0.08% fade per cycle, respectively. The major advantage of the current sol-gel approach is the formation of a porous network structure with uniform particle size by utilizing a carboxylic acid surfactant, which acts as a capping agent preventing and minimizing the agglomeration of the phosphate particles.

2.6. Mechanical Alloying

Recent studies have shown that mechanical alloying or mechanical activation (MA) is a promising method for synthesis of $LiFePO_4$ [77-87], in which the powder particles undergo repeated welding, fracturing and re-welding in a dry high-energy ball-milling vessel. This process results in pulverization and intimate powder mixing. It has been found that a ball-milling step alone is insufficient to obtain a single-phase olivine product. On the other hand, the time and temperature of the thermal treatment necessary for final crystallization of the compound can be decreased substantially by this process [80, 85].

Kim *et al.* [77] prepare olivine $LiFePO_4$ cathode materials by mechanical alloying using iron (Ш) raw material. $LiOH{\cdot}H_2O$, Fe_2O_3, $(NH_4)_2H{\cdot}PO_4$, and acetylene black powders are used as starting materials. The MA process is carried out for 4 h under argon atmosphere using a shaker type ball miller rotating at around 1000 rpm. The mechanical-alloyed powders are then fired from 500 to 900 °C for 30 min in a tube-type vacuum furnace at a pressure 10^{-6} Torr. $LiFePO_4$ synthesized by the above mechanical alloying exhibits excellent cell performance with a discharge capacity of 160 mAh g^{-1}.

Kim *et al.* [79] also report the synthesis of nano-sized $LiFePO_4$ and carbon-coated $LiFePO_4$ ($LiFePO_4/C$) cathode materials by a mechanical activation process. $LiFePO_4$ is synthesized from Li_2CO_3, $FeC_2O_4{\cdot}2H_2O$ and $NH_4H_2PO_4$ taken in stoichiometric quantities. The mechanical activation process consists of the following steps: (i) high-energy ball milling of the powder in a hardened steel vial with zirconia balls at room temperature for different periods in an argon atmosphere using a SPEX mill at 1000 rpm; (ii) conversion of the powder into pellets by mechanical pressing; (iii) thermal treatment of the pellets at temperatures ranging from 500 to 700 °C for different time intervals in a nitrogen atmosphere; (iv) slow cooling to room temperature. $LiFePO_4/C$ with 7.8 wt % acetylene black is prepared by the same processing steps. $LiFePO_4/C$ synthesized by the above mechanical activation process exhibits excellent electrochemical performance, with low capacity fading even at the high current density of 2C.

2.7. Microwave Processing

Microwave processing can achieve very fast and uniform heating through a self-heating process that arises from direct absorption of microwave energy into materials within a short period of time, and at temperatures lower than that required for furnace heating. This processing has been applied in the synthesis of $LiFePO_4$ as a novel heating method [88-93].

Higuchi *et al.* [88] report a novel synthetic method of microwave processing with a domestic microwave oven to prepare $LiFePO_4$ cathode materials. The used starting materials are Li_2CO_3, $NH_4H_2PO_4$, and $Fe(CH_3COO)_2$ or $Fe(CH_3CHOHCOO)_2{\cdot}2H_2O$. These materials are weighed in stoichiometric ratios, dispersed into ethanol, and thoroughly mixed using an agate mortar. The mixed powder is dried at 60 °C and pressed at a pressure of 98 MPa into pellets. Each pellet is covered with glass wool and then placed in an alumina crucible with a lid. The microwave irradiation to the crucible is conducted with a domestic microwave oven that operated at 2.45 GHz, with a maximum power level of 500 W. The charge/discharge result demonstrates that the initial discharge capacity of $LiFePO_4$ synthesized quickly and easily by the above microwave processing is about 125 mAh g^{-1} at 60 °C.

Song *et al.* [89] also demonstrate the synthesis of $LiFePO_4$-C by ball-milling and subsequent microwave heating. Li_3PO_4 and $Fe_3(PO_4)_2{\cdot}8H_2O$ are used as precursor materials. Stoichiometric amounts of Li_3PO_4 and $Fe_3(PO_4)_2{\cdot}8H_2O$ (1:1, molar ratio) are weighed and placed in a ball-milling jar with 5 wt % acetylene black. Ball-milling at various ball-to-powder ratios (weight ratios) is carried out under an Ar atmosphere for 30 min using a vibrant type mill. The ball-milled mixture is pressed into a pellet and then put inside a quartz crucible that is filled with activated carbon. The quartz crucible is put in the middle of a domestic microwave oven (750 W) and microwaves are irradiated for several minutes (2-5 min). During that treatment, carbon generates heat through the direct absorption of microwave energy and thereby makes a reductive atmosphere by carbothermal reaction. The cycling performance demonstrates that $LiFePO_4$-C synthesized by the above ball-milling and subsequent microwave heating can deliver a high initial discharge capacity of 161 mAh g^{-1} at C/10 and exhibit very stable cycling behavior.

2.8. Other Synthesis Methods

Takeuchi *et al.* [94] prepare $LiFePO_4$/C with 20 wt % acetylene black by spark-plasma-sintering process at 600 °C. It is found that $LiFePO_4$ particles are covered with fine carbon particles and they form agglomerates with the size of about 10 μm. The charge/discharge tests for the cell using $LiFePO_4$/C composite positive electrodes show superior cycle performance at the rates of

17-850 mA g^{-1} (1/10-5C) compared with the cell using conventionally blended $LiFePO_4$+C composite positive electrodes. The improvement in the cell performance is attributed to strong binding between $LiFePO_4$ and carbon powders.

Kim *et al.* [95] use $Fe(CH_3COO)_2$, $NH_4H_2PO_4$ and $LiCH_3COO$ as the starting materials to synthesize $LiFePO_4$ by polyol process without any further heating as a post-processing step. The $LiFePO_4$ nanoparticles show a reversible capacity of 166 mAh g^{-1}, which amounts to a utilization efficiency of 98%, with an excellent reversibility in extended cycles.

Wu *et al.* [96] report the synthesis of $LiFePO_4$ by precipitation method. According to the stoichiometry, iron metal, $LiNO_3$, and $(NH_4)_2HPO_4$ are mixed in an aqueous acidic solution. After the starting materials are dissolved, adequate amount of sucrose is added to the solution then heated at 150 °C to evaporate water. The solid residue is calcined at 350 °C for 8 h and then heat-treated at temperatures between 400 and 800 °C for 12 h in N_2. Among the prepared composite cathode materials, the sample heat-treated at 700 °C for 12 h shows better cycling performance than those of others. It shows initial specific discharge capacities of 165 and 130 mAh g^{-1} at 30 °C with C rates of C/10 and 1C, respectively.

Yang *et al.* [97] synthesize small crystallites $LiFePO_4$ powders with conducting carbon coating by ultrasonic spray pyrolysis. The precursor solution for atomization is an aqueous mixing solution of $LiNO_3$, $Fe(NO_3)_3 \cdot 9H_2O$, H_3PO_4, and ascorbic acid ($C_6H_8O_6$) in the de-ionized water at the molar ratio 1:1:1 of Li:Fe:PO_4. The amount of white sugar added into the precursor solution is 60 wt % of $LiFePO_4$ to be formed. The as-sprayed fine powders pyrolysis-synthesized at 450, 550, and 650 °C are heat-treated at 650 °C for 4 h in a tube furnace under a nitrogen atmosphere, and then furnace-cooled to room temperature. The carbon coating on the $LiFePO_4$ surface is critical to the electrochemical performance of $LiFePO_4$ cathode materials of the Li secondary battery, since the carbon coating does not only increase the electronic conductivity via carbon on the surface of particles, but also enhances the ion mobility of Li ion due to prohibiting the grain growth during post-heat-treatment. The carbon of 15 wt % evenly distributed on the final $LiFePO_4$ powders can get the highest initial discharge capacity of 150 mAh g^{-1} at C/10 and 50 °C. Konstantinov *et al.* [98] report the preparation of carbon-mixed $LiFePO_4$ cathode materials by spray solution technology. Ni *et al.* [99] synthesize well-crystallized $LiFePO_4$ by the KCl molten salt method. Lee *et al.* [100, 101] also report the synthesis of $LiFePO_4$ nanoparticles in supercritical water. Carbothermal reduction method [102] and vapor deposition [103] are also utilized to synthesize $LiFePO_4$.

3. How to Improve Electrochemical Performance of $LiFePO_4$ Cathode Materials

3.1. Effect of Particle Size and Morphology on Electrochemical Performance of $LiFePO_4$

For $LiFePO_4$, small particle size and well-shaped crystal are important for enhancing the electrochemical properties [16]. In particles with a small diameter, the Li ions may diffuse over smaller distances between the surfaces and center during Li intercalation and de-intercalation, and $LiFePO_4$ on the particle surfaces contributes mostly to the charge/discharge reaction [45]. This is helpful to enhance the electrochemical properties of $LiFePO_4$/Li batteries because of an increase in the quantity of $LiFePO_4$ particles that can be used.

Many researchers have tried to improve the electrochemical performance by controlling particle size and morphology of $LiFePO_4$ [43-44, 53, 71, 76, 93, 104-115]. Gaberscek *et al.* [107] suggest that based on analysis of nine papers by different authors, the discharge capacity of $LiFePO_4$ drops approximately linearly with average particle size *d*, regardless of the presence/absence of a native carbon coating. Furthermore, the electrode resistance, R_m, as a function of *d*, follows almost exactly the square law: $R_m \propto d^n$ (n = 1.994). Based on theoretical derivation of the same dependence for different contact topologies of interest, they also suggest that the power law with $n = 2$ is generally valid if the low-conductivity species in bulk active particle ($LiFePO_4$) are ions. In particular, to achieve a high-rate capability of $LiFePO_4$, more emphasis should be placed on minimization of *d*, while it is sufficient that the carbon phase or other electronic conductor has only point contacts each individual active particle if the electron-conducting phase also percolates the whole electrode material. In conclusion, they claim that particle size minimization is more important than carbon coating for achieving excellent electrochemical performance.

Liu *et al.* [111] prepare nanocomposites of $LiFePO_4$ with carbon by a solid-state route. Li_2CO_3, $FeC_2O_4 \cdot 2H_2O$, $NH_4H_2PO_4$, and acetylene black as the used starting materials are mixed in ratio of Li : Fe : PO_4 = 1 : 1 : 1 in a planet mixer for 24 h. The mixtures are sintered in a tube furnace at 750 °C for

15 h in an inert atmosphere. The $LiFePO_4/C$ nanocomposites with 5 wt % carbon synthesized by the above solid-state route display d = 100 nm with spherical particle morphology. They suggest that the unique morphology and size are due to admixing of carbon in the starting material, which protects $LiFePO_4$ from oxidation and agglomeration. The cyclic voltammetry results demonstrate that kinetics of Li intercalation and de-intercalation is greatly improved by adding carbon. This amelioration can improve the rate capability of $LiFePO_4/C$.

Ellis *et al.* [53] add the organic additives ascorbic acid and citric acid to the starting materials as carbon sources and reducing agents in the course of $LiFePO_4$ hydrothermal synthesis. They suggest that the size of the crystallites in the absence of organic additives is controlled by the reaction temperature and concentration of the precursors. At 190 °C, typical low concentrations of precursors (7 mmol of $(NH_4)_2Fe(SO_4)_2{\cdot}6H_2O$ in 28 ml of water-or 0.25 M in Fe-along with stoichiometric amounts of H_3PO_4 and $LiOH{\cdot}H_2O$) produce diamond-shaped platelets that are about 250 nm thick. These have large basal dimensions of 1-5 μm. Increasing the reactant concentration by threefold creates more nucleation sites and therefore produces much smaller particles, whose basal size distribution peaks at 250 nm.

The SEM observations of $LiFePO_4$ prepared at low concentration of precursors (0.25 M in Fe) and at 190 °C and subsequent 600 °C confirm that the presence of a reducing agent strongly affects the morphology. The particle size of $LiFePO_4$ prepared from the ascorbic acid is obviously smaller (250-1.5 μm) than that without the reducing agent. Conversely, $LiFePO_4$ prepared from the citric acid contains a wide distribution of particle sizes (500 nm-3 μm), with particle thicknesses remarkably greater than those without additives. The Raman spectrum identifies the deposition of significant quantities of carbon (about 5 wt %) for $LiFePO_4$ prepared from the ascorbic acid. This is possibly because ascorbic acid decomposes near 200 °C under typical conditions. The more stable citric acid does not decompose during the hydrothermal reaction and as a result minimal carbon is detected. These discrepancies in particle size and carbon content are evident in a comparison of the charge/discharge performance of the two materials. With substantially more carbon and smaller average particle size, the $LiFePO_4$ with the ascorbic acid exhibits 70% reversibility on the first cycle, as compared to 35% for the $LiFePO_4$ prepared from citric acid when cycled at a rate of C/10.

Wang *et al.* [105] report the preparation of $LiFePO_4$ via firing amorphous $LiFePO_4$ obtained by chemical reduction and lithiation of $FePO_4$ using Vitamin C (VC) as a reducing agent and Li acetate as Li source in alcohol solution. A solution of precursors is prepared by dissolving 0.06 mol VC and 0.12 mol Li acetate in alcohol, and then 0.1 mol prepared amorphous $FePO_4$ is suspended in the solution. After stirring the suspension at 60 °C for 5 h, the alcohol insoluble amorphous $LiFePO_4$ forms. Crystalline grey $LiFePO_4$ powder is obtained by sintering the amorphous $LiFePO_4$ in furnace at 600 °C for 2 h under Ar (95%) + H_2 atmosphere.

The cycling performance of $LiFePO_4$ prepared by the above non-aqueous method at various charge/discharge rates shows that $LiFePO_4$ exhibits good cycling stability and high reversible capacity. Capacity attenuation is neglectable on cycling. The capacity of $LiFePO_4$ decreases from about 159 mAh g^{-1} at C/10 in the first 45 cycles to about 154 mAh g^{-1}at C/2 rate in the next 10 cycles, and to about 144 mAh g^{-1} at 1C in another 10 cycles and finally recovers to 157 mAh g^{-1} when the discharge rate changes back to C/10. Shortening the diffusion path by synthesizing fine particles is an effective way for improving the high-rate performance of $LiFePO_4$. The ultrafine spherical particles and the conductive carbon between the particles of $LiFePO_4$ are the reasons for its excellent high rate capability.

In addition, Meligrana *et al.* [104] report that $C_{19}H_{42}BrN$ as carbon source and reducing agent can lead to the synthesis of $LiFePO_4$ with finely dispersed nanocrystalline grains. Zaghib *et al.* [113] synthesize $LiFePO_4$ nanoparticles where the size of the particles is small enough that surface effects become important but large enough that their core region is not affected.

3.2. Substitution of Li^+ or Fe^{2+} with Cations

It is known that it is difficult to attain the full capacity because the electronic conductivity of $LiFePO_4$ is very low, which leads to initial capacity loss and poor rate capability, and diffusion of Li^+ ion across the $LiFePO_4/FePO_4$ boundary is slow due to its intrinsic character [16]. Therefore, to improve electrochemical performance of $LiFePO_4$, we should control particle sizes and morphology [43-44, 53, 71, 76, 93, 104-115], as mentioned in section 3.1. Recently, it is found that ionic substitution is another feasible way to enhance the intrinsic electronic conductivity [116-131].

Yamada *et al.* [116-119] report the preparation of Mn-doped $LiMn_{0.6}Fe_{0.4}PO_4$ by solid-state reaction of FeC_2O_4, $MnCO_3$, $NH_4H_2PO_4$, and Li_2CO_3. The used starting materials are dispersed into acetone, then thoroughly mixed, and reground by ball-milling. The mixture is first decomposed at 280 °C and reground again, then heated at 600 °C in purified

N_2 gas flow. The charge/discharge results demonstrate that $LiMn_{0.6}Fe_{0.4}PO_4$ can deliver a discharge capacity of greater than 160 mAh g^{-1}, and $LiMn_{0.6}Fe_{0.4}PO_4$ exhibits two pairs of voltage plateaus, one at 4.1 V (Mn^{3+}/Mn^{2+}) and another at 3.5 V (Fe^{3+}/Fe^{2+}). This is obviously different from the $LiFePO_4$, in which the whole Fe^{3+}/Fe^{2+} reaction proceeds in a two-phase way ($LiFePO_4$-$FePO_4$) with a voltage plateau at 3.4 V [16].

Liu *et al.* [120] synthesize Zn-doped $LiZn_{0.01}Fe_{0.99}PO_4$ by a solid-state reaction. They suggest that the Zn doping promotes the formation of crystal structures, expands the lattice volume and provides more space for lithium-ion intercalation/de-intercalation. In addition, they also claim that the doping decreases the charge transfer resistance, improves the reversibility of lithium-ion intercalation/ de-intercalation, and increases the diffusion of Li ions due to the pillar effect of the doped Zn atoms. The Li ion diffusion coefficient of Zn-doped $LiFePO_4$ increases from 9.98×10^{-14} to 1.58×10^{-13} cm^2 s^{-1}. As results, both discharge capacity and rate capability are greatly ameliorated. After Zn doping, the discharge capacity increases from 88 to 133 mAh g^{-1} at the current density of 0.2 mA cm^{-2} (C/10) in the first cycle.

Wang *et al.* [121] report the preparation of a series of Co-doped $LiFe_{1-x}Co_xPO_4$ solid solutions by solid-state reactions. They suggest that the formation of a solid solution lowers the oxidation potential of the Co^{2+} ions and makes the $Co^{2+}\rightarrow Co^{3+}$ reaction complete at a lower voltage. Consequently, this reaction makes more contribution of capacity in the solid solution than in $LiCoPO_4$. The cycling performance of $LiFe_{1-x}Co_xPO_4$ cycled at a current density of 10 mA g^{-1} demonstrate that both $LiFePO_4$ and $LiCoPO_4$ display the poor cycling performance, only 76.2% and 58.2% the capacity of the first cycle can be retained after 20 cycles for $LiFePO_4$ and $LiCoPO_4$, respectively. Oppositely, $LiFe_{1-x}Co_xPO_4$ solid solutions keep a rather high capacity during 20 cycles, retaining 88.4% of the original capacity for $LiFe_{0.8}Co_{0.2}PO_4$, 86.3% for $LiFe_{0.5}Co_{0.5}PO_4$, and 88.1% for $LiFe_{0.2}Co_{0.8}PO_4$. They claim that electrolyte decomposition should be a reason for the capacity fading of $LiFe_{1-x}Co_xPO_4$ solid solutions as well as for that of $LiCoPO_4$.

Wang *et al.* [122] synthesize $LiFePO_4$ and Ti-doped $LiTi_{0.01}Fe_{0.99}PO_4$ by a sol-gel route. Both $LiFePO_4$ and $LiTi_{0.01}Fe_{0.99}PO_4$ display very flat charge and discharge plateaus. $LiFePO_4$ and $LiTi_{0.01}Fe_{0.99}PO_4$ display initial discharge capacity of 157 and 160 mAh g^{-1} (close to the theoretical capacity of 170 mAh g^{-1}), respectively. They suggest that $LiTi_{0.01}Fe_{0.99}PO_4$ exhibits a slightly higher capacity due to the enhanced electronic conductivity induced by increased p-type semiconductivity through the dopant effect, and a variation of Fe valence during the charging and discharging processes without changing of Fe octahedral coordination symmetry.

Cho *et al.* [123] have examined the effects of La doping on the charge/discharge performance of $LiFe_{0.99}La_{0.01}PO_4/C$ composite cathode materials synthesized by a solid-state reaction. The La doping does not affect the structure of $LiFePO_4$, but remarkably improves its rate capacity performance and cycling stability. They demonstrate that $LiFe_{0.99}La_{0.01}PO_4/C$ can deliver a discharge capacity of 156 mAh g^{-1} cycled in a voltage range of 2.8-4.0 V at C/5, compared to 104 mAh g^{-1} for pure $LiFePO_4$, and sustain 497 cycles based 80% charge retention. They suggest that such a considerable improvement is mainly attributed to enhanced conductivity (from 5.88×10^{-6} to 2.82×10^{-3} S cm^{-1}) and high Li^+ mobility in La-doped $LiFe_{0.99}La_{0.01}PO_4/C$.

Zhang *et al.* [124] report the preparation of $Li_{0.99}Mo_{0.01}FePO_4/C$ composite cathode materials by a solution method followed by calcining at different temperatures. The mix-doping method does not affect the structure of $Li_{0.99}Mo_{0.01}FePO_4/C$ but evidently improves its capacity delivery and cycling performance. They demonstrate that $Li_{0.99}Mo_{0.01}FePO_4/C$ synthesized at 700 °C for 12 h can deliver the initial discharge capacities of 161 and 124 mAh g^{-1} at C/5 and 2C, respectively, which is attributed to the enhanced electronic conductivity by Mo doping and carbon coating. The lower electrochemical polarization of $Li_{0.99}Mo_{0.01}FePO_4/C$ suggests that the enhanced conductivity is induced by the doping method. They claim that two possible conducting mechanisms may be involved. The first probable mechanism, as Chung *et al.* assumed [38], is p-type conduction by the holes generated at the top of the bulk valence Fe–O bands by the activation of the electrons to the empty impurity Mo states. The second probable mechanism is that the doped Mo^{6+}, the vacancies on Li sites, and their neighboring Fe and O ion form a conducting cluster [133]. In addition, the residual carbon resulted from the decomposition of sucrose acts as nucleation site for the formation of $Li_{0.99}Mo_{0.01}FePO_4$ crystals, helping in obtaining samples with uniform sizes. The dispersed carbon particles also promote the electrochemical reaction by enhancing the surface electronic conduction.

According to Ying *et al.* [125], the spherical $Li_{0.97}Cr_{0.01}FePO_4/C$ composites have been synthesized by a controlled crystallization-carbothermal reduction method. They demonstrate that at 0.005, 0.05, 0.1, 0.25 and 1C, $Li_{0.97}Cr_{0.01}FePO_4/C$ can achieve the initial discharge capacity of 163, 151, 142, 131 and 110 mAh g^{-1}, respectively, and also shows excellent cycling performance due to the enhanced electronic conductivity by the Cr^{3+} substitution and carbon coating. The tap-density of the spherical

$Li_{0.97}Cr_{0.01}FePO_4/C$ powders is as high as 1.8 g cm^{-3}, which is greatly higher than the non-spherical $LiFePO_4$ powders reported. They claim that the high-density spherical $Li_{0.97}Cr_{0.01}FePO_4/C$ cathode materials can provide significant incentive for battery manufactures to consider it as a very promising candidate to be utilized in the lithium-ion batteries with high power density.

Hong *et al.* [126] synthesize $LiFe_{0.9}Mg_{0.1}PO_4$ by mechanical alloying method followed by heat treatments. The prepared $LiFe_{0.9}Mg_{0.1}PO_4$ shows an equilibrium potential plateau in two-phase region with a potential hysteresis of 18 mV between Li insertion and extraction, and has a high rate capability. Due to the fast charge-transfer reaction, high electronic and ionic diffusivity, the phase transformation between $LiFe_{0.9}Mg_{0.1}PO_4$ and $Fe_{0.9}Mg_{0.1}PO_4$ begins to play an important role in the charge/discharge process.

In addition, the improved electrochemical performances of $LiM_xFe_{1-x}PO_4$ and $Li_{1-x}M_xFePO_4$ (Ti, Zr, Mg) [127], $Li_{0.98}Al_{0.02}FePO_4/C$ [128], $Li_{0.99}Ti_{0.01}FePO_4/C$ [129], $LiFe_{0.9}M_{0.1}PO_4$ (M = Ni, Co, Mg) [130-131], and $Li_{0.99}Al_{0.01}FePO_4/C$ [132] are also reported.

3.3. Effect of Carbon Coating and Metal or Metal Oxide Mixing on Charge/Discharge Performance of $LiFePO_4$

It is well-known that carbon as a reducing agent can not only prevent the formation of Fe^{3+} impurity and the agglomeration of particles during the preparation of $LiFePO_4$, but also increase the electronic conductivity.

Ravet *et al.* [134] are the first to show that carbon-coated $LiFePO_4$ with 1 wt % carbon can deliver a discharge capacity of 160 mAh g^{-1} at 80 °C at a discharge rate of C/10 using a polymer electrolyte.

Huang *et al.* [135] have made a systematic study of nanocomposites of $LiFePO_4$ and conductive carbon by two different methods. Method A employs a composite of $LiFePO_4$ with a carbon xerogel formed from a resorcinol-formaldehyde precursor; method B uses surface-oxidized carbon particles to act as a nucleating agent for $LiFePO_4$ growth. Both particle size minimization and intimate carbon contact are necessary to optimize electrochemical performance. The resultant $LiFePO_4/C$ composite using method A can deliver 90% theoretical capacity at C/2, with very good rate capability and excellent stability.

Prosini *et al.* [136] synthesize $LiFePO_4$ by the solid-state reaction of Li_2CO_3, $FeC_2O_4{\cdot}H_2O$ and $(NH_4)_2HPO_4$ in the presence of high-surface area carbon-black. The SEM observations demonstrate that the adding of the fine carbon powders reduces $LiFePO_4$ grain size. The carbon is evenly dispersed among grains, ensuring a good electric contact. $LiFePO_4$ composite cathode materials are conductive and no additional carbon-black has to be added during the electrode preparation. Thus, the electrochemical properties of $LiFePO_4$ are greatly improved. $LiFePO_4$ composite cathode materials can achieve a discharge capacity of 125 mAh g^{-1} at a discharge rate of C/10. The discharge capacity increases with temperatures and the full discharge capacity can be obtained at 80 °C and C/10 discharge rate. $LiFePO_4$ composite cathode materials may be cycled 230 times at C/2 discharge rate and room temperature, delivering an average discharge capacity of 95 mAh g^{-1}, with a very satisfactory discharge capacity retention.

Shin *et al.* [83] have investigated the electrochemical performance of carbon-coated $LiFePO_4$ using three different carbon sources such as graphite, carbon black, and acetylene black. The SEM observations reveal that the carbon-coated $LiFePO_4$ consists of non-uniform fine particles with the size range of 100-300 nm, which are much smaller than the pure $LiFePO_4$ particles. This implies that the presence of carbon in the mixture retards the particle growth during calcining. The electronic conductivities of the carbon-coated $LiFePO_4$ are 10^{-2}-10^{-4} S cm^{-1}, which are much higher than 10^{-9}-10^{-10} S cm^{-1} of $LiFePO_4$. They suggest that this improvement is attributed to the excellent electrical contacts between $LiFePO_4$ particles by the carbon layer. Thus, the electrochemical performance of the carbon-coated $LiFePO_4$ shows higher discharge capacity and better capacity retention compared to $LiFePO_4$. $LiFePO_4$ coated with graphite exhibits better electrochemical performance than others. The carbon-coated $LiFePO_4$ can deliver a discharge capacity of 120 mAh g^{-1} at 2C and room temperature. Equivalent circuit analysis from impedance measurement confirms that the improved electrochemical performance of the carbon-coated $LiFePO_4$ using graphite is induced by the low charge transfer resistance and low Li-ion migration resistance.

Thorat *et al.* [137] describe the preparation and testing of $LiFePO_4$ cathodes for hybrid vehicle application. $LiFePO_4$ cathodes contain combinations of three different carbon conductivity additives: vapor-grown carbon fibers (CF), carbon black (CB) and graphite (GR). SEM observations reveal that $LiFePO_4$ cathodes containing carbon fibers (CB+CF and CF only) show the fibers quite clearly. The fibers appear to be in good contact with other particles. The fibers are believed to improve the electrical conduction and contact throughout the cathode and also provide mechanical strength to the solid matrix. They suggest that the combination of fibers and carbon black can provide a highly conductive network that connects well to the active material particles and the current collector. $LiFePO_4$ cathodes with a mixture of CF+CB exhibits the best power-performance, followed by cells containing CF only and then by CB+GR. The improved electrode performance due to the fibers also allows an increase in energy density while still meeting power goals. The best specific-power performance for each of the compositions investigated occurs around an active material loading of 1 mAh cm^{-2}. The maximum discharge rate that leads to 2.2 V at the end of the pulse is about 20.6C, obtained by interpolation. The specific power corresponding to the maximum rate is 3882 W kg^{-1} cathode, again obtained by interpolation.

With the exclusion of carbon black, graphite, acetylene black and vapor-grown carbon fibers as carbon conductive additives, multiwalled carbon nanotubes (MWCNTs) are also used as a carbon conductive additive. MWCNTs have many merits over amorphous acetylene black, such as high conductivity, small specific surface area and tubular shape. Thess *et al.* [138] report that electronic conductivity of MWCNTs thin film is about $(1\text{-}4)\times10^2$ S cm^{-1} along the nanotube axis and 5-25 S cm^{-1} perpendicular to the axis, respectively.

Li *et al.* [139] have studied $LiFePO_4$/MWCNTs novel network composite cathode compared to $LiFePO_4$/acetylene black cathode. The SEM observations reveal that a piece of MWCNTs connect $LiFePO_4$ particles in series and countless MWCNTs interlace all particles together to form a three-dimensional network wiring, the electron conducting on the interface between cathode particles and current collector is greatly improved when MWCNTs act as a conducting bridge. The charge/discharge testing results demonstrate that MWCNTs can improve cycling efficiency and rate capability more effectively on the same conditions than carbon black. A variety of oxo-functional groups may exist on the surface of acetylene black. These external functional groups and micropores on the surface contribute to the irreversible reactions with electrolytes [140]. However, MWCNTs can prevent these irreversible reactions and improve cycling efficiency due to deletion of oxides groups and reduction of specific surface area. $LiFePO_4$/MWCNTs composite cathode materials can achieve the initial discharge capacities of 155 mAh g^{-1} at C/10 and 146 mAh g^{-1} at 1C rate.

We also study the electrochemical performance of $LiFePO_4$/MWCNTs composite cathode materials synthesized by a hydrothermal method in lithium polymer batteries. The SEM observations show that the MWCNTs intertwine with $LiFePO_4$ particles together to form a three-dimensional network. The dispersed MWCNTs provide pathways for electron transference. Therefore, the electronic conductivity of $LiFePO_4$-MWCNTs composites is improved.

The electronic conductivities are 5.86×10^{-9} S cm^{-1} for pure $LiFePO_4$, 1.08×10^{-1} S cm^{-1} for $LiFePO_4$-MWCNTs with 5 wt % MWCNTs. Figure 4 shows the cyclic voltammograms of $LiFePO_4$-MWCNTs with different MWCNTs contents at a scan rate of 0.1 mV s^{-1}. It can be seen that the redox peak profile of $LiFePO_4$-MWCNTs with 5 wt % MWCNTs is more symmetric and spiculate than that of $LiFePO_4$, demonstrating that the reversibility and reactivity of $LiFePO_4$-MWCNTs with 5 wt % MWCNTs are enhanced due to improvement of electronic conductivity and the fast ionic diffusion kinetics resulting from a decrease in the crystallite size by MWCNTs. As shown in Figure 5, the discharge rate capability of $LiFePO_4$-MWCNTs with 5 wt % MWCNTs is obviously ameliorated by MWCNTs. $LiFePO_4$-MWCNTs with 5 wt % MWCNTs can deliver the discharge capacities of 123 mAh g^{-1} at C/10, 110 mAh g^{-1} at 3C/10, 106 mAh g^{-1} at C/2, 97 mAh g^{-1} at 1C and 53 mAh g^{-1} at 3C.

Spong *et al.* [141] report the preparation of carbon-coated $LiFePO_4$ by a novel, one-step, low-cost synthesis method from aqueous precursor solutions of $Fe(NO_3)_3$, $LiCH_3COO$, H_3PO_4 and sucrose. Sucrose additions up to a mole fraction of 25% are found to suppress crystallization of the salts during the first stages of pyrolysis, thereby reducing elemental segregation and facilitating the formation of the olivine structure below 500 °C in a single heating step. Sucrose also acts as a reducing agent and a source of carbon to form a conductive network in the active material during synthesis, leading to a higher capacity than materials in which sucrose is substituted with acetylene black. After additional treatment with sucrose at 700 °C, carbon-coated $LiFePO_4$ can achieve the discharge capacities of 162 mAh g^{-1} at C/14 rate and 158 mAh g^{-1} at C/3.5 in the voltage range of 2.0-4.5 V.

Yun *et al.* [41] use poly(vinyl alcohol) (PVA) as a carbon source to prepare $LiFePO_4$/C composite cathode materials by a conventional solid-state reaction with one-step heat treatment at 800 °C. They show that carbon coating can control particle growth, provide improved electrical contact between particles, and enhance the surface electronic conductivity—all of which improve electrochemical performance, especially rate capacity. The charge/discharge testing results indicate that $LiFePO_4$/C composite cathode material with 5 wt % PVA exhibits the best electrochemical performance, and can deliver a discharge capacity of 153 mAh g^{-1} at C/10 with excellent capacity retention.

In addition to the above carbon sources, there are still naphthalenetetracarboxylic dianhydride [142], hydroxyethyl-cellulose [70], white table sugar [143], polypropylene [144], propylene [103], glycol [145],

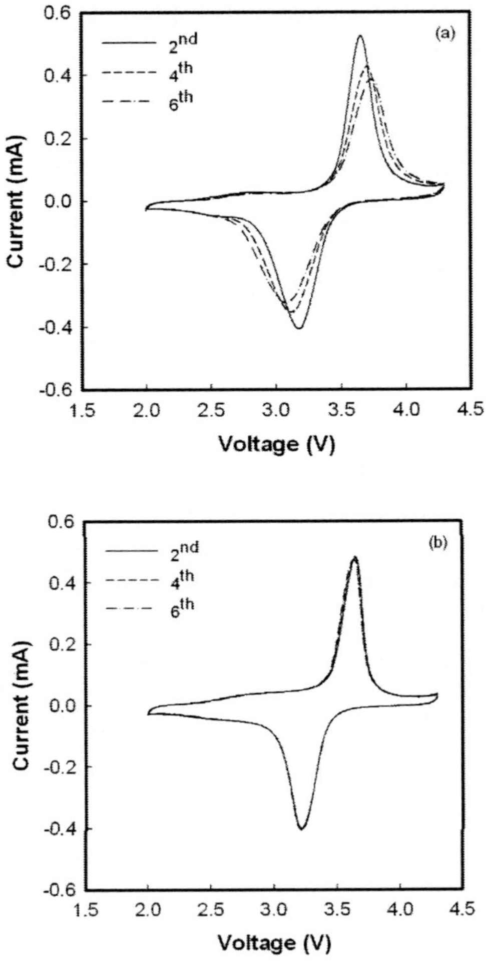

Figure 4. The cyclic voltammograms of $LiFePO_4$-MWCNTs with: (a) 0 wt %, and (b) 5 wt % MWCNTs at a scan rate of 0.1 mV s^{-1}.

citric acid monohydrate [146] and kitchen oils (olive, soybean and butter) [147] for the preparation of $LiFePO_4$/C composite cathode materials.

Croce *et al.* [148] report the preparation and electrochemical performance of kinetically improved Cu-added or Ag-added $LiFePO_4$ composite cathode materials. The added Cu or Ag metal powders do not affect the structure of $LiFePO_4$ but clearly improve its kinetics in terms of capacity delivery and cycling life due to a reduction of the particle size and an increase of the bulk intra- and inter-particle electronic conductivity of $LiFePO_4$. The obvious capacity improvement of Ag-added $LiFePO_4$ both at medium (C/5) and particularly at high (1C) rates is maintained for many cycles, demonstrating the stability of Ag-added $LiFePO_4$.

According to Liu *et al.* [149], ZrO_2 nanolayer coated $LiFePO_4$ particles have been successfully synthesized by a chemical precipitation method. The HR-TEM observations reveal that nanolayer structured ZrO_2 with a thickness of 2-3 nm exists on the surface of $LiFePO_4$ particles. The ZrO_2 nanolayer increases the mechanical toughness of the core particles and decreases the interface charge transfer resistance. It does not affect the crystal structure of $LiFePO_4$ core but considerably improves the electrochemical properties at high charge/discharge rate due to the amelioration of the electrochemical dynamics on the $LiFePO_4$ electrode/electrolyte interface. Furthermore, the ZrO_2 nanolayer is favorable to increasing the thermal stability by forming a more stable solid electrolyte interface layer and covering the over-reactive sites on the particle surface to avoid probable electrolyte decomposition. In addition, the ZrO_2 surface coating can also provide a protective layer for $LiFePO_4$ core particles to shield them from direct contact with the acidic electrolyte. ZrO_2 nanolayer coated $LiFePO_4$ can deliver the initial discharge capacities of 146 mAh g^{-1} at C/10 and 97 mAh g^{-1} at 1C with excellent capacity retention.

In addition, the enhanced electrochemical properties of ZnO-coated $LiFePO_4$ [150], $LiFePO_4$-Ag composite thin films [151] and polypyrrole-added $LiFePO_4$ composites [152] are also reported.

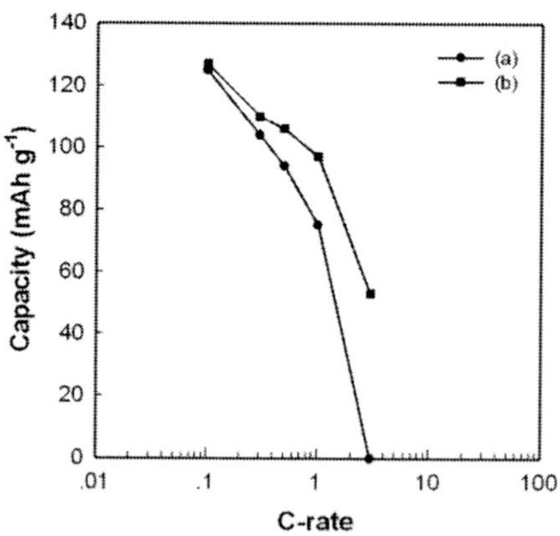

Figure 5. The rate capability of $LiFePO_4$-MWCNTs with: (a) 0 wt %, and (b) 5 wt % MWCNTs at various C rates ranging from C/10 to 3C rate at room temperature.

4. Summary and Future Prospect

$LiFePO_4$ cathode materials have been reviewed focusing mainly on the synthesis method and how to improve the electrochemical performance. For $LiFePO_4$, small particle size and well-shaped crystals are important for enhancing the electrochemical properties [16]. In particles with a small diameter, the Li ions may diffuse over shorter distances between the surfaces and center during Li intercalation and de-intercalation, and the $LiFePO_4$ on the particle surfaces contributes mostly to the charge/discharge reaction [45]. This is helpful to enhance the electrochemical properties of $LiFePO_4$/Li batteries because of an increase in the quantity of $LiFePO_4$ particles that can be used. Among the various synthesis methods as mentioned above, the hydrothermal synthesis is a useful method to prepare fine particles, and has some advantages such as simple synthesis process, and low energy consumption, compared to high firing temperature and long firing time during solid-state reaction used conventionally.

Although $LiFePO_4$ possesses high stability, low cost and high compatibility with environment, it suffers from the limitations of poor electronic conductivity and slow Li-ion diffusion, and therefore operates unsatisfactorily at lower temperatures and/or higher current densities. Coating $LiFePO_4$ active particles with conductive carbon [83], carbon mixing as a powder initially [136] and in-situ generation by organic compounds during the preparation [145] is a feasible method to overcome its insulating nature and make the cell operate at high current densities.

These continuous effects to improve the synthesis method and the electrochemical performance of $LiFePO_4$ will result in Li-ion batteries with higher energy density and lower price, and larger scale applications including low current density applications, such as mobile phones, laptop computers and digital cameras, and high current density applications, such as electrical vehicles and hybrid electrical vehicles.

5. ACKNOWLEDGMENTS

The authors acknowledge the financial supports from National Key Basic Research and Development Program of China (Grant No. 2010CB631001) and the China Postdoctoral Science Foundation Funded Project (Grant No. 20090451124).

REFERENCES

[1] Ozawa, K. *Solid State Ion*., 1994, 69, 212.

[2] Pistoia, G; Zane, D; Zhang, Y. *J. Electrochem. Soc*., 1995, 142, 2551.

[3] Resimers, JN; Dahn, JR; Sacken, U von. *J. Electrochem. Soc*., 1993, 140, 2752.

[4] Li, W; Resimers, JN; Dahn, JR. *Solid State Ion*., 1993, 67, 123.

[5] Dahn, JR; Sacken, U von; Juzkow, MW; Al-Janaby, H. *J. Electrochem. Soc*., 1991, 138, 2207.

[6] Koetschau, I; Richard, MN; Dahn, JR; Soupart, JB; Rousche, JC. *J. Electrochem. Soc*., 1995, 142, 2906.

[7] Jeong, IS; Kim, JU; Gu, HB. *J. Power Sources*., 2001, 102, 55.

[8] Jin, B; Kim, JU; Gu, HB. *J. Power Sources*., 2003, 117, 148.

[9] Kim, JU; Jo, YJ; Park, GC; Gu, HB. *J. Power Sources*., 2003, 119-121, 686.

[10] Gu, YX; Chen, DR; Jiao, XL. *J. Phys. Chem. B*., 2005, 109, 17901.

[11] Sauvage, F; Tarascon, JM; Baudrin, E. *J. Phys. Chem.C*., 2007, 111, 9264.

[12] Zheng, HH; Zhang, HC; Fu, YB; Abe, T; Ogumi, Z. *J. Phys. Chem. B*., 2005, 109, 13676.

[13] Luo, JY; Wang, YG; Xiong, HM; Xia, YY. *Chem. Mater*., 2007, 19, 4791.

[14] Lee, HC; Chang, SK; Goh, EY; Jeong, JY; Lee, JH; Kim, HJ; Cho, JJ; Hong, ST. *Chem. Mater*., 2008, 20, 5.

[15] Jayalakshmi, M; Rao, MM; Scholz, F. *Langmuir*., 2003, 19, 8403.

[16] Padhi, AK; Nanjundaswamy, KS; Goodenough, JB. *J. Electrochem. Soc*., 1997, 144, 1188.

[17] Bramnik, NN; Bramnik, KG; Buhrmester, T; Baehtz, C; Ehrenberg, H; Fuess, H. *J. Solid State Electrochem*., 2004, 8, 558.

[18] Jin, B; Gu, HB; Kim, KW. *J. Solid State Electrochem*., 2008, 12, 105.

[19] Amine, K; Yasuda, H; Yamachi, M. *Electrochem. Solid State Lett*., 2000, 3, 178.

[20] Okada, S; Sawa, S; Egashira, M; Yamaki, J; Tabuchi, M; Kageyama, H; Konishi, T; Yoshino, A. *J. Power Sources*., 2001, 97-98, 430.

[21] Yamada, A; Hosoya, M; Chung, SC; Kudo, Y; Hinokuma, K; Liu, KY; Nishi, Y. *J. Power Sources*., 2003, 119-121, 232.

[22] Deniard, P; Dulac, AM; Rocquefelte, X; Grigorova, V; Lebacq, O; Pasturel, A; Jobic, S. *J. Phys. Chem. Solids*., 2004, 65, 229.

[23] Loris, JM; Perez-Vicente, C; Tirado, JL. *Electrochem. Solid State Lett*.,

2002, 5, A234.
[24] Li, G; Azuma, H; Tohda, M. *Electrochem. Solid State Lett.*, 2002, 5, A135.
[25] Gu, HB; Jin, B; Jun, DK; Han, ZJ. *J. Nanosci. Nanotechnol.*, 2007, 7, 4037.
[26] Gu, HB; Jun, DK; Park, GC; Jin, B; Jin, EM. *J. Nanosci. Nanotechnol.*, 2007, 7, 3980.
[27] Hu, YS; Guo, YG; Dominko R; Gaberscek M; Jamnik J; Maier J. *Adv. Mater.*, 2007, 19, 1963.
[28] Xie, HM; Wang, RS; Ying, JR; Zhang, LY; Jalbout, AF; Yu, HY; Yang GL; Pan, XM; Su, ZM. *Adv. Mater.*, 2006, 18, 2609.
[29] Wang, YQ; Wang, JL; Yang, J; Nuli, YN. *Adv. Funct. Mater.*, 2006, 16, 2135.
[30] Fisher, CAJ; Prieto, VMH; Islam, MS. *Chem. Mater.*, 2008, 20, 5907.
[31] Delacourt, C; Poizot, P; Morcrette, M; Tarascon, JM; Masquelier, C. *Chem. Mater.*, 2004, 16, 93.
[32] Wang, LN; Li, ZC; Xu, HJ; Zhang, KL. *J. Phys. Chem. C.*, 2008, 112, 308.
[33] Bramnik, NN; Nikolowski, K; Baehtz, C; Bramnik, KG; Ehrenberg, H. *Chem. Mater.*, 2007, 19, 908.
[34] Zaghib, K; Mauger, A; Goodenough, JB; Gendron, F; Julien, CM. *Chem. Mater.*, 2007, 19, 3740.
[35] Islam, MS; Driscoll, DJ; Fisher, CAJ; Slater, PR. *Chem. Mater.*, 2005, 17, 5085.
[36] Shiraishi, K; Dokko, K; Kanamura, K. *J. Power Sources.*, 2005, 146, 555.
[37] Myung, ST; Komaba, S; Hirosaki, N; Yashiro, H; Kumagai, N. *Electrochim. Acta.*, 2004, 49, 4213.
[38] Chung, SY; Bloking, JT; Chiang, YM. *Nat. Mater.*, 2002, 1, 123.
[39] Molenda, J; Stoklosa, A; Bak, T. *Solid State Ion.*, 1989, 36, 53.
[40] Shimakawas, Y; Numata, T; Tabuchi, J. *J. Solid State Chem.*, 1997, 131, 138.
[41] Yun, NJ; Ha, HW; Jeong, KH; Park, HY; Kim, K. *J. Power Sources.*, 2006, 160, 1361.
[42] Gabrisch, H; Wilcox, JD; Doeff, MM. *Electrochem. Solid-State Lett.*, 2006, 9, A360.
[43] Yamada, A; Chung, SC; Hinikuma, K. *J. Electrochem. Soc.*, 2001, 148, A224.

[44] Gibot, P; Cabanas, MC; Laffont, L; Levasseur, S; Carlach, P; Hamelet, S; Tarascon, JM; Masquelier, C. *Nat. Mater.*, 2008, 7, 741.
[45] Takahashi, M; Tobishima, S; Takei, K; Sakurai, Y. *J. Power Sources.*, 2001, 97-98, 508.
[46] Yang, SF; Song, YN; Ngala, K; Zavalij, PY; Whittingham, MS. *J. Power Sources.*, 2003, 119-121, 239.
[47] Kim, HS; Cho, BW; Cho, WI. *J. Power Sources.*, 2004, 132, 235.
[48] Andersson, AS; Thomas, JO. *J. Power Sources.*, 2001, 97-98, 498.
[49] Kim, DK; Park, HM; Jung, SJ; Jeong, YU; Lee, JH; Kim, JJ. *J. Power Sources.*, 2006, 159, 237.
[50] Yang, S; Zavalij, PY; Whittingham, MS. *Electrochem. Commun.*, 2001, 3, 505.
[51] Dokko, K; Shiraishi, K; Kanamura, K. *J. Electrochem. Soc.*, 2005, 152, A2199.
[52] Chen, JJ; Whittingham, MS. *Electrochem. Commun.*, 2006, 8, 855.
[53] Ellis, B; Kan, WH; Makahnouk, WRM; Nazar, LF. *J. Mater. Chem.*, 2007, 17, 3248.
[54] Chen, JJ; Vacchio, MJ; Wang, SJ; Chernova, N; Zavalij, PY; Whittingham, MS. *Solid State Ion.*, 2008, 178, 1676.
[55] Dokko, K; Koizumi, S; Nakano, H; Kanamura, K. *J. Mater. Chem.*, 2007, 17, 4803.
[56] Chen, G; Song, X; Richardson, TJ. *Electrochem. Solid-State Lett.*, 2006, 9, A295.
[57] Jin, B; Jin, EM; Park, KH; Gu, HB. *Electrochem. Commun.*, 2008, 10, 1537.
[58] Jin, B; Gu, HB. *Solid State Ion.*, 2008, 178, 1907.
[59] Jin, B; Gu, HB; Zhang, WX; Park, KH; Sun, GP. *J. Solid State Electrochem.*, 2008, 12, 1549.
[60] Jin, EM; Jin, B; Jun, DK; Park, KH; Gu, HB; Kim, KW. *J. Power Sources.*, 2008, 178, 801.
[61] Yang, MR; Ke, WH; Wu, SH. *J. Power Sources.*, 2005, 146, 539.
[62] Park, K; Kang, K; Lee, S; Kim, G; Park, Y; Kim, H. *Mater. Res. Bull.*, 2004, 39, 1803.
[63] Arnold, A; Garche, J; Hemmer, R; Ströbele, Vogler, C; Wohlfahrt-Mehrens, M. *J. Power Sources.*, 2003, 119-121, 247.
[64] Ni, JF; Zhou, HH; Chen, JT; Zhang, XX. *Mater. Lett.*, 2005, 59, 2361.
[65] Park, K; Son, J; Chung, H; Kim, S; Lee, C; Kim, H. *Electrochem. Commun.*, 2003, 5, 839.
[66] Prosini, PP; Carewska, M; Scaccia, S; Wisniewski, P; Passerini, S;

Pasquali, M. *J. Electrochem. Soc.*, 2002, 149, A886.
[67] Franger, S; Le Cras, F. Bourbon, C; Rouault, H. *Electrochem. Solid-State Lett.*, 2002, 5, A231.
[68] Cho, TH; Chung, HT. *J. Power Sources.*, 2004, 133, 272.
[69] Dominko, R; Goupil, JM; Bele, M; Gaberscek, M; Remskar, M; Hanzel, D; Jamnik, J. *J. Electrochem. Soc.*, 2005, 152, A858.
[70] Dominko, R; Bele, M; Gaberscek, M; Remskar, M; Hanzel, D; Pejovnik, S; Jamnik, J. *J. Electrochem. Soc.*, 2005, 152, A607.
[71] Choi, DW; Kumta, PN. *J. Power Sources.*, 2007, 163, 1064.
[72] Sanchez, M; Brito, G; Fantini, M; Goya, G; Matos, J. *Solid State Ion.*, 2006, 177, 497.
[73] Gaberscek, M; Dominko, R; Bele, M; Remskar, M; Hanzel, D; Jamnik, J. *Solid State Ion.*, 2005, 176, 1801.
[74] Xu, Z; Xu, L; Lai, Q; Ji, X. *Mater. Res. Bull.*, 2007, 42, 883.
[75] Wilcox, JD; Doeff, MM; Marcinek, M; Kostecki, R. *J. Electrochem. Soc.*, 2007, 154, A389.
[76] Gabrisch, H; Wilcox, JD; Doeff, MM. *Electrochem. Solid-State Lett.*, 2008, 11, A25.
[77] Kim, CW; Lee, MH; Jeong, WT; Lee, KS. *J. Power Sources.*, 2005, 146, 534.
[78] Kim, CW; Park, JS; Lee, KS. *J. Power Sources.*, 2006, 163, 144.
[79] Kim, JK; Cheruvally, G; Choi, JW; Kim, JUk; Ahn, JH; Cho, GB; Kim, KW; Ahn, HJ. *J. Power Sources.*, 2007, 166, 211.
[80] Kwon, SJ; Kim, CW; Jeong, WT; Lee, KS. *J. Power Sources.*, 2004, 137, 93.
[81] Liao, XZ; Ma, ZF; Wang, L; Zhang, XM; Jiang, Y; He, YS. *Electrochem. Solid-State Lett.*, 2004, 7, A522.
[82] Shin, HC; Chung, KY; Min, WS; Byun, DJ; Jang, H; Cho, BW. *Electrochem. Commun.*, 2008, 10, 536.
[83] Shin, HC; Cho, WI; Jang, H. *Electrochim. Acta.*, 2006, 52, 1472.
[84] Shin, HC; Cho, WI; Jang, H. *J. Power Sources.*, 2006, 159, 1383.
[85] Franger, S; Bourbon, C; Cras, FL. *J. Electrochem. Soc.*, 2004, 151, A1024.
[86] Franger, S; Cras, FL; Bourbon, C; Rouault, H. *J. Power Sources.*, 2003, 119-121, 252.
[87] Kosova, N; Devyatkina, E. *Solid State Ion.*, 2004, 172, 181.
[88] Higuchi, M; Katayama, K; Azuma, Y; Yukawa, M; Suhara, M. *J. Power Sources.*, 2003, 119-121, 258.

[89] Song, MM; Kang, YM; Kim, JH; Kim, HS; Kim, DY; Kwon, HS; Lee, JY. *J. Power Sources*., 2007, 166, 260.
[90] Park, KS; Son, JT; Chung, HT; Kim, SJ; Lee, CH; Kim, HG. *Electrochem. Commun*., 2003, 5, 839.
[91] Wang, XJ; Ren, JX; Li, YZ; Wei, JP; Gao, XP; Yan, J. *Chin. J. Inorg. Chem*., 2005, 21, 249.
[92] Nakayama, M; Watanabe, K; Ikuta, H; Uchimoto, Y; Wakihara, M. *Solid State Ion*., 2003, 164, 35.
[93] Wang, L; Huang, YD; Jiang, RR; Jia, DZ *Electrochim. Acta*., 2007, 52, 6778.
[94] Takeuchi, T; Tabuchi, M; Nakashima, A; Nakamura, T; Miwa, Y; Kageyama, H; Tatsumi, K. *J. Power Sources*., 2005, 146, 575.
[95] Kim, DH; Kim, J. *Electrochem. Solid-State Lett*., 2006, 9, A439.
[96] Wu, SH; Hsiao, KM; Liu, WR. *J. Power Sources*., 2005, 146, 550.
[97] Yang, MR; Teng, T, H; Wu, SH. *J. Power Sources*., 2006, 159, 307.

[98] Konstantinov, K; Bewlay, S; Wang, G; Lindsay, M; Wang, J; Liu, H; Dou, S; Ahn, J. *Electrochim. Acta*., 2004, 50, 421.
[99] Ni, JF; Zhou, HH; Chen, JT; Zhang, XX. *Mater. Lett*., 2007, 61, 1260.
[100] Lee, J; Teja, AS. *Mater. Lett*., 2006, 60, 2105.
[101] Lee, J; Teja, AS. *J. Supercrit. Fluid*., 2005, 35, 83.
[102] Barker, J; Saidi, MY; Swoyer, JL. *J. Electrochem. Soc*., 2003, 6, A53.
[103] Belharouak, I; Johnson, C; Amine, K. *Electrochem. Commun*., 2005, 7, 983.
[104] Meligrana, G; Gerbaldi, C; Tuel, A; Bodoardo, S; Penazzi, N. *J. Power Sources*., 2006, 160, 516.
[105] Wang, BF; Qiu, YL; Ni, SY. *Solid State Ion*., 2007, 178, 843.
[106] Xia, YG; Yoshio, M; Noguchi, H. *Electrochim. Acta*., 2006, 52, 240.
[107] Wang, L; Zhang, Z; Zhang, K. *J. Power Sources*., 2007, 167, 200.
[108] Kim, DH; Kim, J. *J. Phys. Chem. Solids*., 2007, 68, 734.
[109] Gaberscek, M; Dominko, R; Jamnik, J. *Electrochem. Commun*., 2007, 9, 2778.
[110] Kobayashi, Nishimura, SI; Park, MS; Kanno, R; Yashima, M; Ida, T; Yamada, A. *Adv. Funct. Mater*., 2009, 19, 395.
[111] Liu, H; Li, C; Zhang, HP; Fu, LJ; Wu, YP; Wu, HQ. *J. Power Sources*., 2006, 159, 717.
[112] Salah, AA; Mauger, A; Zaghib, K; Goodenough, JB; Ravet, N; Gauthier, M; Gendron, F; Julien, CM. *J. Electrochem. Soc*., 2006, 153, A1692.
[113] Zaghib, K; Mauger, A; Gemdron, F; Julien, CM. *Chem. Mater*., 2008,

20, 462.
[114] Sides, CR; Croce, F; Young, VY; Martin, CR; Scrosati, B. *Electrochem. Solid-State Lett.*, 2005, 8, A484.
[115] Delacourt, C; Poizot, P; Levasseur, S; Masquelier, C. *Electrochem. Solid-State Lett.*, 2006, 9, A352.
[116] Yamada, A; Chung, SC. *J. Electrochem. Soc.*, 2001, 148, A960.
[117] Yamada, A; Kudo, Y; Liu, KY. *J. Electrochem. Soc.*, 2001, 148, A1153.
[118] Yamada, A; Kudo, Y; Liu, KY. *J. Electrochem. Soc.*, 2001, 148, A747.
[119] Yamada, A; Hosoya, M; Chung, SC; Kudo, Y; Hinokuma, K; Liu, KY; Nishi, Y. *J. Power Sources.*, 2003, 119-121, 232.
[120] Liu, H; Cao, Q; Fu, LJ; Li, C; Wu, YP; Wu, HQ. *Electrochem. Commun.*, 2006, 8, 1553.
[121] Wang, DY; Wang, ZX; Huang, XJ; Chen, LQ. *J. Power Sources.*, 2005, 146, 580.
[122] Wang, GX; Bewwlay, S; Needham, SA; Liu, HK; Liu, RS; Drozd, VA; Lee, JF; Chen, JM. *J. Electrochem. Soc.*, 2006, 153, A25.

[123] Cho, YD; Fey, GTK; Kao, HM. *J. Solid State Electrochem.*, 2008, 12, 815.
[124] Zhang, M; Jiao, LF; Yuan, HT; Wang, YM; Guo, J; Zhao, M; Wang, W; Zhou, XD. *Solid State Ion.*, 2006, 177, 3309.
[125] Ying, JR; Lei, M; Jiang, CY; Wan, CR; He, XM; Li, JJ; Wang, L; Ren, JG. *J. Power Sources.*, 2006, 158, 543.
[126] Hong, J; Wang, CS; Kasavajjula, U. *J. Power Sources.*, 2006, 162, 1289.
[127] Wang, GX; Needham, SA; Yao, J; Wang, JZ; Liu, RS; Liu, HK. *J. Power Sources.*, 2006, 159, 282.
[128] Xie, H; Zhou, ZT. *Electrochim. Acta.*, 2006, 51, 2063.
[129] Wang, G; Cheng, Y; Yan, MM; Jiang, ZY. *J. Solid State Electrochem.*, 2007, 11, 457.
[130] Wang, DY; Li, H; Shi, SQ; Huang, XJ; Chen, LQ. *Electrochim. Acta.*, 2005, 50, 2955.
[131] Wang, GX; Bewlay, SL; Konstantinov, K; Liu, HK; Dou, SX; Ahn, JH. *Electrochim. Acta.*, 2004, 50, 443.
[132] Hsu, KF; Tsay, SY; Hwang, BJ. *J. Power Sources.*, 2005, 146, 529.
[133] Shi, SQ; Liu, LJ; Yang, CY; Wang, DS; Wang, ZX; Chen, LQ; Huang, XJ. *Phys. Rev. B.*, 2003, 68, 195108.
[134] Ravet, N; Goodenough, JB; Besner, S; Simoneau, M; Hovington, P; Armand, M. *Proceedings of the 196th ECS Meeting.*, 1999, 99-102, Abst. 127.

[135] Huang, H; Yin, SC; Nazar, LF. *Electrochem. Solid-State Lett.*, 2001, 4, A170.

[136] Prosini, PP; Zane, D; Pasquali, M. *Electrochim. Acta.*, 2001, 46, 3517.

[137] Thorat, IV; Mathur, V; Harb, JN; Wheeler, DT. *J. Power Sources.*, 2006, 162, 673.

[138] Thess, A; Lee, R; Nikolaev, P. *Science.*, 1996, 273, 483.

[139] Li, XL; Kang, FY; Bai, XD; Shen, WC. *Electrochem. Commun.*, 2007, 9, 663.

[140] Mukai, SR; Hasegawa, T; Takagi, M; Tamon, H. *Carbon.*, 2004, 42, 837.

[141] Spong, AD; Vitins, G; Owen, JR. *J. Electrochem. Soc.*, 2005, 152, A2376.

[142] Doeff, MM; Hu, Y; McLarnon, F; Kostecki, R. *Electrochem. Solid-State Lett.*, 2003, 6, A207.

[143] Chen, Z; Dahn, JR. *J. Electrochem. Soc.*, 2002, 149, A1184.

[144] Mi, CH; Zhao, XB; Cao, GS; Tu, JP. *J. Electrochem. Soc.*, 2005, 152, A483.

[145] Wang, BF; Qiu, YL; Yang, L. *Electrochem. Commun.*, 2006, 8, 1801.

[146] Palomares, V; Goni, A; Muro, IGD; Meatza, ID; Bengoechea, M; Miguel, O; Rojo, T. *J. Power Sources.*, 2007, 171, 879.

[147] Kim. K; Jeong, JH; Kim, IJ; Kim, HS. *J. Power Sources.*, 2007, 167, 524.

[148] Croce, F; Epifanio, AD; Hassoun, J; Deptula, A; Olczac, T; Scrosati, B. *Electrochem. Solid-State Lett.*, 2002, 5, A47.

[149] Liu, H; Wang, GX; Wexler, D; Wang, JZ; Liu, HK. *Electrochem. Commun.*, 2008, 10, 165.

[150] Leon, B; Vicente, CP; Tirado, JL; Biensan, P; Tessier, C. *J. Electrochem. Soc.*, 2008, 155, A211.

[151] Lu, ZG; Cheng, H; Lo, MF; Chung, CY. *Adv. Funct. Mater.*, 2007, 17, 3885.

[152] Wang, GX; Yang, L; Chen, Y; Wang, JZ; Bewlay, S; Liu, HK. *Electrochim. Acta.*, 2005, 50, 4649.

In: Lithium Batteries: Research, Technology... ISBN: 978-1-60741-722-4
Editors: G. R. Dahlin and K. E. Strom

Chapter 2

INORGANIC CATHODE MATERIALS FOR LITHIUM ION BATTERIES

Zhicong Shi [a]***, Hansan Liu*** [b] ***and Jiujun Zhang*** [b]

a. State Key Laboratory of Fine Chemicals, Dalian University of Technology P.O. BOX 132, 158-Zhongshan Road, Dalian, Liaoning, 116012, China

b. Institute for Fuel Cell Innovation, National Research Council of Canada 4250 Wesbrook Mall, Vancouver, BC, V6T 1W5, Canada

1. INTRODUCTION

Lithium ion batteries, a class of chemical power sources that use an electrochemical process of lithium ion intercalation into or de-intercalation from host materials, are gaining dominance in mobile electronic applications, and also showing promise for an upcoming new generation of electric vehicle applications. Currently, the most successful active electrode materials used in lithium ion batteries are graphite (anode material with a specific capacity of 350 mA h g^{-1}) and $LiCoO_2$ (cathode material with a specific capacity of 135 mA h g^{-1}). Under the driving force of safety issues, a new cathode material, $LiFePO_4$, has been developed in recent years as the most promising cathode material for next-generation lithium ion batteries. However, this new material can deliver a specific capacity of only 150 mA h g^{-1}, which is far less than that of anode materials (Figure 1) [[2]]. The low specific capacity of cathode materials has been identified as the factor preventing lithium ion batteries from meeting the high capacity and high power demands of automobiles and electronic devices. Therefore, finding cathode materials with higher specific capacities has become the key priority in lithium ion battery research and development (R&D).

In general, a cathode material for lithium ion batteries needs to meet the following requirements [[3]]:

(1) The material can react with lithium reversibly and remain a stable structure during the process of intercalation/de-intercalation. This requirement is essential for extending the lifetime of lithium ion batteries.
(2) The reaction free energy of the cathode material with lithium must be high enough to achieve a battery with high energy density.
(3) The material must have high electronic conductivity and high lithium ion conductivity to facilitate fast charge transferring and then deliver a high power density.
(4) The material does not chemically react with the electrolyte during cycling. This is a basic safety requirement.
(5) The material is low-cost and environmentally friendly.

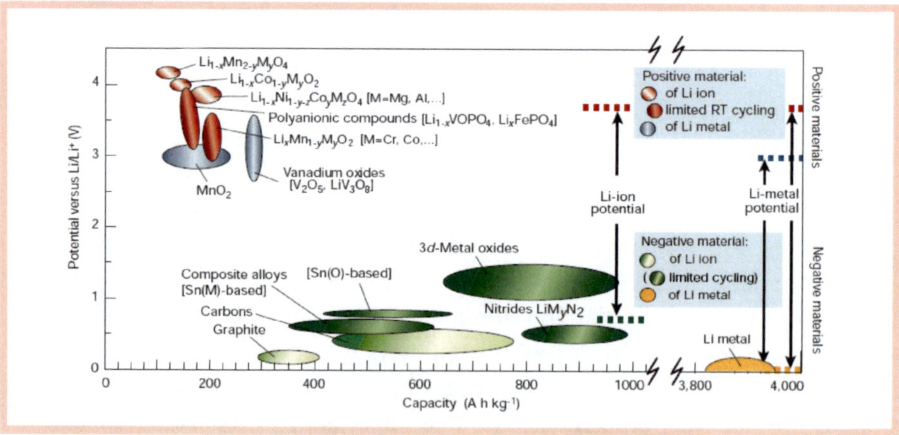

Figure 1. Voltage versus capacity for cathode and anode materials presently used or under serious considerations for next generation of rechargeable Li-based batteries. Note the big difference in capacity between cathode and anode materials, which is the reason why cathode material is the bottle-neck of capacity density of lithium ion batteries.[1]

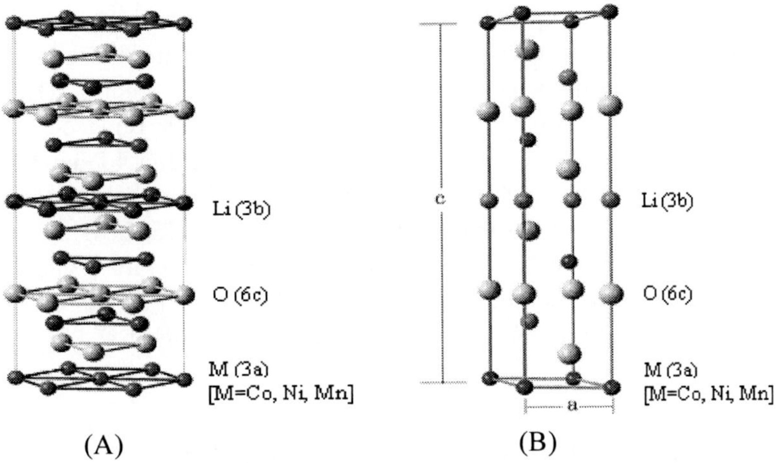

Figure 2. (A) Ball-stick structure model of hexagonal layered structure $LiMO_2$ (M = Mn, Co, or Ni) and (B) unit cell of $LiMO_2$ (M = Mn, Co, or Ni).

Two general classes of cathode material are candidates for lithium ion batteries: inorganic compounds and organic polymers. The most popular inorganic cathode materials can be divided into three kinds of inorganic metal compounds. The first consists of the lithium transition metal oxides with a layered α-$NaFeO_2$ structure (such as $LiCoO_2$, $LiNiO_2$, and $LiMnO_2$); the second comprises the lithium transition metal oxides with a spinel structure (such as $LiMn_2O_4$); and the third is the group of lithium transition metal phosphates (polyanionic compounds) with an olivine structure (such as $LiFePO_4$ and $LiMnPO_4$) or with a NASICON structure (such as $Li_3V_2(PO_4)_3$). Three decades ago, Goodenough and his team [4] found that $LiCoO_2$ had a layered α-$NaFeO_2$ structure and could electrochemically release lithium ions during a battery reaction, suggesting that this material could be used as a cathode material for lithium ion batteries. In 1991, SONY successfully employed this kind of material as the cathode in their first commercialized lithium ion batteries, and opened a new era of rechargeable batteries [5]. The layered α-$NaFeO_2$ structure of $LiCoO_2$ has a cubic close-packed (ccp) oxygen lattice. Complete removal of lithium ions from the interslab can result in a rearrangement of the oxygen lattice into a hexagonal close-packed (hcp) frame [6]. A stable delithiated structure can only be obtained by 50% lithium removal, which limits the maximum practical specific capacity of $LiCoO_2$ to 135 mA h g^{-1}. Besides this moderate specific capacity, $LiCoO_2$ also has the disadvantages of being unsafe, toxic, and costly. Although lattice doping and surface coating may improve its lifetime and safety [7] [8], $LiCoO_2$ is considered a less than ideal lithium ion battery cathode material for large-scale applications such as power sources in hybrid electric vehicles (HEV) and electric vehicles (EV). An alternative cathode is expected to replace $LiCoO_2$.

In this chapter, we review most of the efforts made to develop new inorganic cathode materials for a new generation of lithium ion batteries. We focus primarily on layered $LiNiO_2$ and $LiMnO_2$, spinel $LiMn_2O_4$, as well as olivine $LiFePO_4$, presenting in detail their crystal structures, intercalation mechanisms, synthesis methods, and performance. Future R&D directions and potential applications of these cathode materials are also discussed.

2. Layered Lithium Metal Oxides

2.1 Introduction

An ideal lithium metal oxide $LiMO_2$ (M = Mn, Co, or Ni) has an α-$NaFeO_2$ rock-salt structure with a space group of R-3m (No. 166), as shown in Figure 2. The atomic coordinates are regulated as M at the 3a site (0,0,0), Li at the 3b site (0,0,0.5), and O at the 6c site (0,0,z) (0,0,-z). The value of z is around 0.25, with small deviations dependent on the property of the transition metal M. The oxygen sub-lattice in the rock-salt structure takes an ABCABCABC... stacking sequence. The cations occupy the octahedral sites of alternating layers parallel to the crystal plane (111), thus yielding a structure of AγBaCβAcBαC (Greek letters denote transition metal layers and small Latin letters denote Li layers) (Figure 2). Therefore, a two-dimensional path on an a×b panel can facilitate the diffusion of lithium ions during intercalation or extraction.

Structural changes in the cathode material during battery reactions can affect battery lifetime. For example, it has been recognized that removal of Li ions during the charging process may change the phase structure of $Li_{1-x}MO_2$ due to distortion of the ccp oxygen lattice. In particular, layered metal oxides might be partially changed to an energetically favorable spinel structure when a composition of $Li_{0.5}MO_2$ is reached. The deterioration of crystal structure from layered to spinel during electrochemical cycling was previously observed for both $LiCoO_2$ and $LiMnO_2$, using transmission electron microscopy (TEM) [9] [10], micro-Raman [11], and neutron diffraction coupled with nuclear magnetic resonance (NMR) [12]. This structural transfer could change both electrochemical activity and cathode lifetime [13]. Therefore, to achieve deep removal and re-insertion of lithium ions for higher energy density and longer battery lifetime, improved structural stabilization of layered $LiMO_2$ is necessary.

2.2 $LiNiO_2$

2.2.1 Problems with $LiNiO_2$

$LiNiO_2$ has an α-$NaFeO_2$ structure with a space group of R-3m (No. 166), which is the same as that in $LiCoO_2$ (Figure 2). The theoretical specific capacity of $LiNiO_2$ is as high as 276 mA h g^{-1}, and the material structure can remain stable even when the Li is removed at a Li/Ni ratio of 0.65 when the battery is electrochemically cycling between 2.5 V and 4.1 V. This can result in $LiNiO_2$ having a practical specific capacity of 180 mA h g^{-1}, which is higher than $LiCoO_2$ (135 mA h g^{-1}) [14]. However, $LiNiO_2$ has some limitations as a cathode material in lithium ion batteries, despite nickel being more readily available than cobalt [3]. Firstly, enough excess nickel exists in the lithium layer to form a non-stoichiometric $[Li^{+}_{1-z}Ni^{2+}_{z}]_{3b}[Ni^{3+}_{1-z}Ni^{2+}_{z}]_{3a}[O_2]_{6c}$, which could block the lithium diffusion route and thus reduce the lithium diffusion coefficient. The non-stoichiometric $[Li^{+}_{1-z}Ni^{2+}_{z}]_{3b}[Ni^{3+}_{1-z}Ni^{2+}_{z}]_{3a}[O_2]_{6c}$ can also seriously limit the power capability of the $LiNiO_2$. Unfortunately, it is difficult to eliminate this undesired excess nickel from the material synthesis process. Secondly, phase transformation of $LiNiO_2$ during lithium extraction/re-insertion cycles can cause an irreversible change in crystal structure, leading to a short cycling lifetime. Finally, the delithiated Li_xNiO_2 has high oxidization potential for reaction with the organic solvent electrolyte, causing battery safety issues. These three challenges must be overcome before $LiNiO_2$-based materials can be used as cathodes in commercial lithium ion batteries.

2.2.2 Synthesis of stoichiometric $LiNiO_2$-based materials

Some general difficulties arise in the synthesis of stoichiometric $LiNiO_2$ using the traditional solid-state method. This is due to: (1) defects on lithium sites when lithium is evaporated at high temperatures [15]; (2) the large energy barrier for the oxidation of Ni^{2+} to Ni^{3+} [16]; and (3) the decomposition and phase transformation of $LiNiO_2$ at high temperatures [17]. Nickel on lithium sites in non-stoichiometric $[Li^{+}_{1-z}Ni^{2+}_{z}]_{3b}[Ni^{3+}_{1-z}Ni^{2+}_{z}]_{3a}[O_2]_{6c}$ can cause a large capacity loss in the first charge/discharge cycle, followed by poor capacity upon further cycling [18].

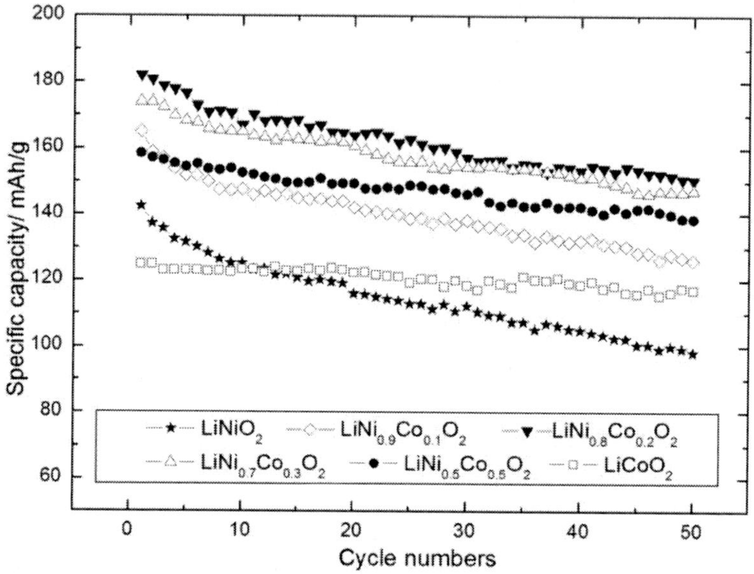

Figure 3. Specific discharge capacities during charge-discharge cycling of $LiNi_{1-y}Co_yO_2$ (y=0, 0.1, 0.2, 0.3, 0.5, 1.0) with a discharge current density of 18mA g^{-1} (0.1C) between 3.0V and 4.2V. [Hansan Liu's unpublished result]

In terms of mitigation, three strategies can improve the synthesis of a near-stoichiometric $LiNiO_2$- based material: (1) using excess lithium salt in the reaction precursor to compensate for lithium evaporation at high temperatures; (2) using a low-temperature method, such as sol-gel or co-precipitation, in place of the conventional solid-state method; and (3) doping with a second metal, such as $LiNi_{1-y}Co_yO_2(0 \leqslant y \leqslant 1)$ solid solution, to reduce atomic displacement and then enhance the ordering of the hexagonal layered structure [19].

In our previous work [26, 27, 37], we studied the synthesis, structures, and performance of $LiNi_{1-y}Co_yO_2$ (y = 0, 0.1, 0.2, 0.3, 0.5, 1.0) as cathode materials. A sol-gel method using citric acid as a chelating reagent was developed for preparing the materials at a relative low temperature (725 °C) and in a short time (24 hrs). All of the above three strategies were adopted in this synthesis. The crystal structures of the materials were analyzed by the Rietveld refinement method based on their X-ray diffraction data. The result indicated that the ordering of the hexagonal layered structure was enhanced by cobalt content. For pure $LiNiO_2$, 7.3% of the Li 3b sites were occupied by nickel. The non-stoichiometric number was decreased to 2.4% after 20% of the nickel was replaced by cobalt, and almost no nickel could be found on Li 3a sites when the cobalt doping level was increased to 30%. Cobalt doping also had significant effect on cathode performance. As presented in Figure 3, pure $LiNiO_2$ showed an initial specific discharge capacity of 143 mA h g^{-1}, higher than that of $LiCoO_2$ (125 mA h g^{-1}), under a discharge current density of 18 mA g^{-1} (0.1 C) between 3.0 V and 4.2 V. Unfortunately, its capacity retention was far poorer than that of $LiCoO_2$; only 68% of the initial capacity was retained after 50 cycles. With cobalt doping, the specific capacity of $LiNi_{0.8}Co_{0.2}O_2$ increased to 185 mA h g^{-1}, and 82% of the initial specific capacity was retained after 50 cycles. However, when the cobalt doping level exceeded 20%, the excess doping suppressed removal of lithium from the layered structure, resulting in a lower reversible capacity. The optimal doping content was determined to be ~20% cobalt.

2.2.3 Structural stability of delithiated $LiNiO_2$-based materials

Irreversible crystal structure change is the main reason for capacity fading during charge/discharge cycling of $LiNiO_2$-based materials. An ideal host material has a stable crystal structure or undergoes reversible structural change during charge/discharge cycling. It is well known that most oxides are formed by electrovalent bonding and interaction between ions, and their crystal structures therefore must change during ion insertion or extraction. Ohzuku et al. [[20]] and Delmas et al. [[21]] carried out in-depth studies of structural changes in $LiNiO_2$ during the charge/discharge processes. After the removal of lithium ions, the crystal structure of Li_xNiO_2 was transformed from a rhombohedral phase (R1, 1.00>x>0.75) to a monoclinic phase (M, 0.75>x>0.45), then a new rhombohedral phase (R2, 0.45>x>0.25), followed by a third rhombohedral phase (R3, 0.25>x>0.00), and finally a hexagonal phase (H4, x = 0) (Figure 4a). This successive phase transformation was believed to be due to the Jahn-Teller effect of the NiO_6 octahedron, and the rearrangement of the super crystal structure formed by lithium/vacancy ordering during lithium ion removal and hole generation [[22]].

The transformation of multiple phases of $LiNiO_2$ during lithium intercalation and extraction can cause serious capacity fading. The partial irreversible phase transformation and the crystal cell volume change results in the cracking and break-off of active materials. Moreover, the high oxidation state of nickel in Li_xNiO_2 can also lead to a reaction with the ethyl carbonate (EC) in the organic electrolyte when the charging voltage reaches 4.3 V (Figure 4a), which could result in a large irreversible capacity and cause serious safety problems [[23]]. Therefore, enhancing structural stability during cycling is one of the major challenges in the commercialization of $LiNiO_2$-based cathode materials.

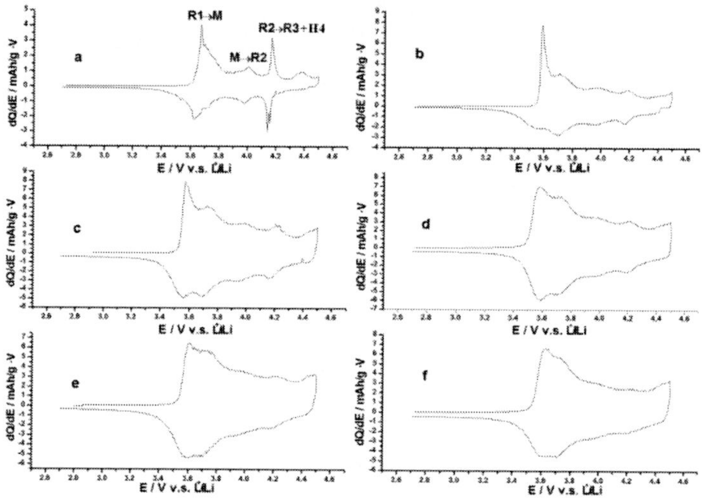

Figure 4. Differential capacity (dQ/dE) against voltage curves derived from the second charge-discharge cycle (0.2C, 2.7-4.5V) of $LiNiO_2$(a), $LiNi_{0.8}Co_{0.2}O_2$(b) and $LiNi_{0.8-y}Ti_yCo_{0.2}O_2$ (y=0.025(c), 0.050(d), 0.075(e), 0.100(f)). [Hansan Liu's unpublished result]

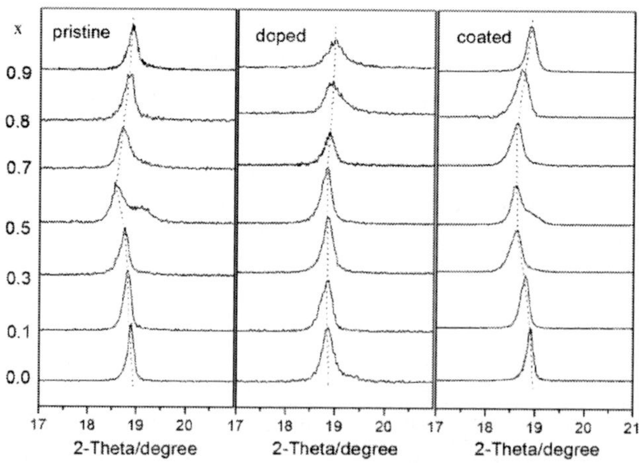

Figure 5. The (003) diffraction peaks of ex-situ XRD patterns for the delithiated pristine, 5% Ti-doped and 3% TiO_2-coated $Li_{1-x}Ni_{0.8}Co_{0.2}O_2$ at different lithium content during charge/discharge process. [37]

Doping with other metal elements is a demonstrated successful strategy to improve the structural stability of $LiNiO_2$-based cathode materials. Many studies have deployed this doping strategy by replacing part of the nickel with other metal elements such as cobalt, magnesium, aluminum, or titanium [24] [25] [26] [27]. It is well known that cobalt doping in $LiNiO_2$ can reduce disorder in the interslab and improve its structure stability. The weaker redox peak at high voltage in Figure 4b supports this assertion. When $LiNiO_2$ is doped by redox inactive metal ions, such as the titanium in $LiNi_{0.8-y}Ti_yCo_{0.2}O_2$, complete removal of the lithium can be avoided, thereby stabilizing the crystal structure and suppressing the irreversible phase changes that occur with very low lithium content (Figure 4c–f). This doping effect can be demonstrated by ex-situ X-ray diffraction (XRD) of delithiated cathode materials $Li_{1-x}Ni_{0.8-y}Ti_yCo_{0.2}O_2$, as shown in Figure 5 [28]. A series of enlarged (003) diffraction peaks, which reflect the interslab distances, are extracted from the ex-situ XRD patterns of the delithiated cathode materials, including $Li_{1-x}Ni_{0.8-y}Ti_yCo_{0.2}O_2$. For a delithiated pristine material, the (003) peak shifts slightly toward the lower diffraction angles when the lithium content x increases to 0.5 (Figure 5, left pattern). Further lithium extraction can make the (003) peak shift slightly back to higher diffraction angles. The (003) peak shift corresponds to the reciprocating changes in interslab distance, i.e., structural instability during the delithiation process. Compared to pristine materials, the materials doped with 5% Ti show better structure stability during the delithiation process. As shown in the central pattern of Figure 5, before $x = 0.5$, there is no significant shift for the (003) peak. Even when $x > 0.5$, only a slight shift to higher angles is observed.

Coating is another strategy for improving structural stability. Figure 5 can be used to compare the doping effect and the coating effect. Figure 5 (right pattern) shows the structural change of a 3% TiO_2-coated $Li_{1-x}Ni_{0.8}Co_{0.2}O_2$ material after delithiation. It can be seen that a similar change in the (003) peak as that of pristine material occurred in the coated materials, indicating that there still were obvious structural changes during cycling for the coated material. Therefore, the doping strategy is better than the coating strategy in terms of improving structural stability.

The effect of doping on structural stability can also lead to better capacity retention and longer cycling lifetime for $LiNi_{0.8-y}Ti_yCo_{0.2}O_2$ cathode materials. For example, $LiNi_{0.7}Ti_{0.1}Co_{0.2}O_2$ can give an initial specific discharge capacity of 188 mA h g^{-1} in the range of 2.7–4.5 V and fade from 188 to 148 mA h g^{-1} after 100 cycles with a high capacity retention of 80%, which is much better than the 58% retention for $LiNi_{0.8}Co_{0.2}O_2$ cathode material (Figure 6) [27]. If another metal element was doped to form a quaternary lithium metal oxide, the cyclicability of the lithium ion battery could be further improved. For example, additional Mg-doping, yielding $LiNi_{0.7}Co_{0.2}Ti_{0.05}Mg_{0.05}O_2$, could result in an improved capacity retention of up to 91% after 100 cycles [29].

2.2.4 Thermal stability of delithiated $LiNiO_2$-based materials

Poor thermal stability of Li_xNiO_2 materials in a charged state, caused by self-decomposition of nickel oxide and the high oxidization ability of Ni^{4+} with an organic solvent electrolyte, is another major factor that degrades capacity during charge/discharge cycling. This poor thermal stability also makes $LiNiO_2$-based batteries unsafe. Generally speaking, cathode materials show good thermal stability at full lithiation, but would decompose at low temperature and low lithium content state. Differential scanning calorimetry (DSC) showed the thermal stability order of three oxide cathodes when charged to 4.2 V in the same electrolyte (PC/EC/DMC(1/1/3)+$LiPF_6$(1M)) to be as follows: $LiMn_2O_4$ > $LiCoO_2$ > $LiNiO_2$ [30]. $LiNiO_2$ could decompose to NiO at 850 °C [31], while $Li_{0.3}NiO_2$ could decompose at 200 °C with a specific thermal capacity of 1600 J g^{-1} [32] [33]. Cathode decomposition releases large quantities of heat and gas, causing a fatal blast during the lithium ion battery operation.

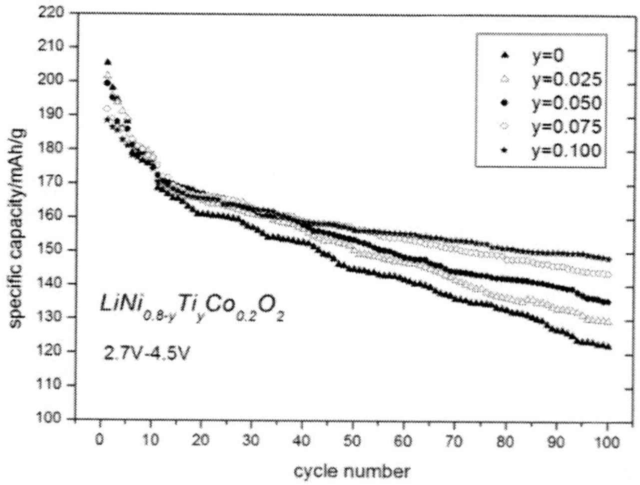

Figure 6. Plots of discharge-specific capacity vs. cycle number for $LiNi_{0.8-y}Ti_yCo_{0.2}O_2$ (y = 0, 0.025, 0.050, 0.075, 0.100) cathode materials. Cycling was carried out with a current rate of 0. 2C for 1-5 cycles, 0.5C for 6–10 cycles and 1C for 11-100 cycles. [26]

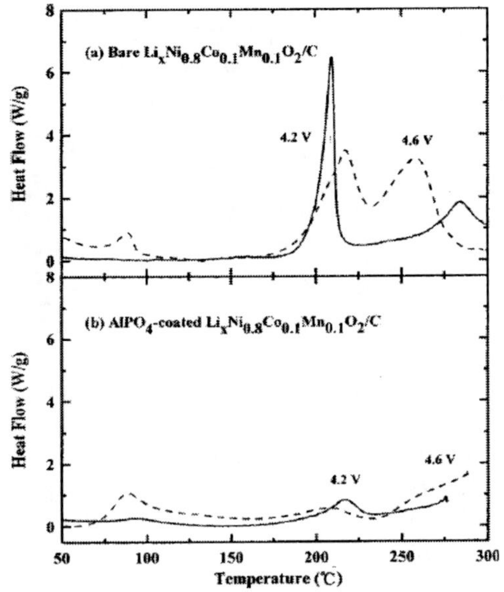

Figure 7. DSC scans of the bare and AlPO4-coated cathodes at 4.2 and 4.6 V vs. carbon (~4.3 and ~4.7 V vs. lithium, respectively). The cathodes were extracted from the Li-ion cells, and the scan rate was 3 °C min^{-1}. [36]

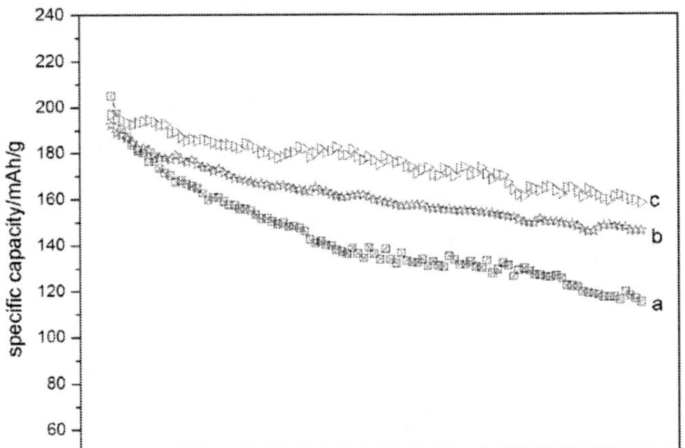

Figure 8. Cycling stability curves of (a) pristine, (b) 5% Ti-doped and (c) 3% TiO2-coated $LiNi_{0.8}Co_{0.2}O_2$ at 0.2C current density between 2.7V and 4.5V. [37]

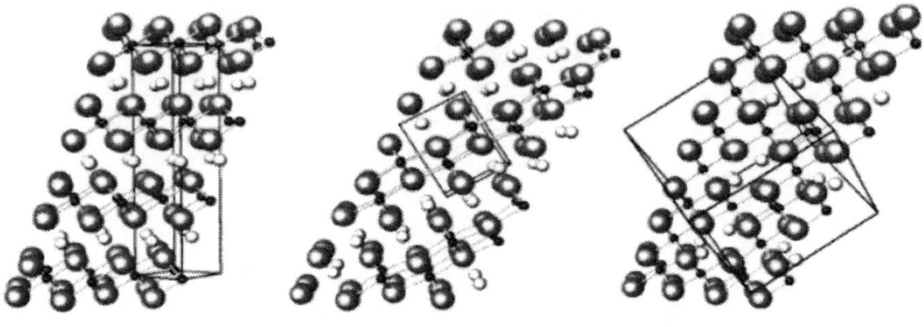

a b c

Figure 9. Crystal structures of a) R-3m layered $LiMO_2$, b) Pmnm o- $LiMnO_2$, c) Fd3m spinel $LiMn_2O_4$. Small white spheres: Li. Small black spheres: Transition metal. Large gray spheres: Oxygen. Note the similar oxygen sublattices in all the structure. Layered $LiMO_2$ has alternating layers of Li and M. *o*- $LiMO_2$ shows zig-zag layering. Spinel has alternating layers filled 3/4 and 1/4 by Mn, resulting in three-dimensional channels with Li in tetrahedral sites. [38]

Two strategies exist to mitigate the effect of thermal instability: (1) doping with other metal elements, and (2) coating with thermally stable oxides or phosphates. As shown in section 2.2.3, doping with cobalt and titanium can improve the thermal stability of $LiNiO_2$ cathode materials, as demonstrated by the shift in the electrolyte's anodic peak from 4.35 V in pure $LiNiO_2$ to a higher voltage in the doped samples (Figure 4). On the other hand, surface coating using chemically stable metal oxides or metal phosphates has been demonstrated to be effective in preventing a direct reaction between the oxidative component and the organic electrolyte, and then in reducing the thermal effect on lithium ion batteries during cycling [34] [35] [36] [37]. For example, coating $AlPO_4$ on $Li_xNi_{0.8}Co_{0.1}Mn_{0.1}O_2$ could lead to a significantly reduced exothermic heat, just one-quarter of the heat released by bare $LiNiO_2$ at charged states (Figure 7) [37]. It has been found that if $LiNi_{0.8}Co_{0.2}O_2$ was coated with a 15-nm layer of TiO_2, the anodic peak of electrolyte oxidation on the cyclic voltammetry (CV) curves was also suppressed [38]. As shown in Figure 5, the ex-situ XRD pattern shows no difference in crystal structure evolution between pristine and TiO_2-coated $LiNi_{0.8}Co_{0.2}O_2$ during the charge/discharge process. This means that surface coating has no effect on the structural stability of the cathode material during cycling. However, the discharge capacity retention can be improved, as shown in Figure 8, resulting in extended lithium ion battery lifetime. This positive effect should be attributed to the enhanced thermal stability of $LiNi_{0.8}Co_{0.2}O_2$ in the delithiated state.

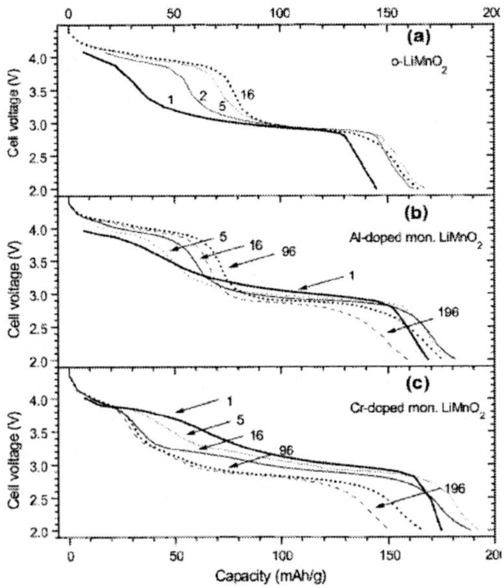

Figure 10. Discharge voltage profiles for (a) o-$LiMnO_2$ (b) 5% Al-doped m-$LiMnO_2$, and (c) 3% Cr-doped m-$LiMnO_2$ during extended cycling of Li cells at 55 °C. Current rate is 30 mA g^{-1}. [46]

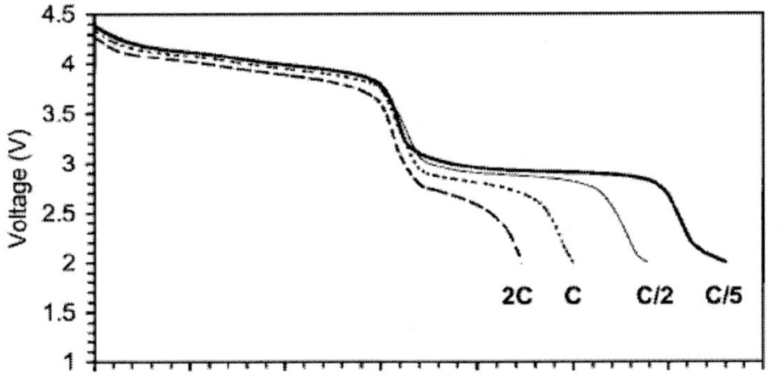

Figure 11. Discharge voltage profiles for orthorhombic $LiMnO_2$ prepared at high temperature in a Li cell discharged at C/5, C/2,C, and 2C rates, ambient temperature. It shows the typical discharge behavior of spinel phase with 4V and 2.8V plateaus. [38]

2.3 $LiMnO_2$

2.3.1 Challenges of $LiMnO_2$

Mn-based cathodes, primarily layered $LiMnO_2$ and spinel $LiMn_2O_4$, are very interesting in their application to large batteries, because they are superior to lithium cobalt or nickel oxides in terms of safety, cost, and toxicity. $LiMnO_2$ has a high theoretical discharge capacity of 285 mA h g^{-1}, about twice that of $LiMn_2O_4$. In comparison with both hexagonal $LiCoO_2$ and $LiNiO_2$, $LiMnO_2$ does not have a perfect α-$NaFeO_2$ structure (Figure 9a) [39]. The trivalent Mn ions can cause a cooperative distortion of the MnO_6 octahedra due to Jahn-Teller stabilization, leading to a metastable monoclinic unit cell (space group C2/m), denoted as m-$LiMnO_2$. This new structure shows a lower symmetry with different angles and lattice constants, compared to higher rhombohedral symmetry. The thermodynamically stable $LiMnO_2$ has an orthorhombic symmetry, denoted as *o*-$LiMnO_2$ (Figure 9b). However, both m-$LiMnO_2$ and *o*-$LiMnO_2$ have the cation ordering of a layered α-$NaFeO_2$ structure, which tends to gradually transform into a spinel structure (Figure 9c) during lithium intercalation/de-intercalation. Phase transformation often happens among the orthorhombic, the layered O_3, and the spinel structures due to the same close-packed oxygen sub-lattice and only minor differences in cation occupation (Figure 9). Energetically, the spinel structure is preferable over the layered O_3 or orthorhombic structure for most $Li_{0.5}MO_2$ [40]. A recent study using ab initio calculations showed that the delithiated Li_xMnO_2 layered materials could transform to a spinel structure in a two-stage process [41] [42]. In the first stage, part of the Mn and Li ions rapidly migrated into tetrahedral sites surrounded by Li vacancies. The activation barrier to the migration of Mn into a tetrahedral site was low, partly because of the charge disproportionation of Mn^{3+} into Mn^{2+} (tetrahedral) and Mn^{4+} (octahedral). In the second stage, the structural transformation involved a more difficult coordinated rearrangement of Mn and Li ions to form spinels, which took place more slowly due to its complexity and higher activation barriers. This phase transformation could cause fast capacity fading and reduce the lifetime of layered $LiMnO_2$ cathodes. Therefore, the challenge for practical application of $LiMnO_2$ cathodes is how to prepare and stabilize the structure of $LiMnO_2$ during cycling.

2.3.2 Development of monoclinic $LiMnO_2$ cathode materials

Much effort has been put into developing the commercial and scientific potential of layered $LiMnO_2$. Layered $LiMnO_2$ materials do not crystallize in the R-3m space group, but show a monoclinic distortion of the lattice (space group C2/m) due to the cooperative ordering of Jahn-Teller distorted [$Mn^{3+}O_6$] octahedra. Although m-$LiMnO_2$ is thermodynamically metastable when compared to both orthorhombic and spinel phases, it has been successfully synthesized by soft chemical methods such as ion exchange and hydrothermal synthesis.

The exchange of Na^+ ions in layered $NaMnO_2$ with Li^+ ions was carried out to form m-$LiMnO_2$ [42-44]. An earlier attempt at Li-Na exchange in molten salts failed when the layered structure collapsed [43]. Using a modified ion exchange strategy, a m-$LiMnO_2$ material was successfully prepared [44] [45]. m-$LiMnO_2$ was first achieved by ion exchange of layered $NaMnO_2$ in methanol with LiCl at 90 °C. However, the ion-exchange kinetics was so poor that the process took approximately a month [44]. The ion exchange process was then accelerated by refluxing layered $NaMnO_2$ in n-hexanol with LiBr at 150 °C [45]. Unfortunately, the m-$LiMnO_2$ material did not exhibit good electrochemical performance. A large amount of lithium could be extracted on the first charge, but the ions could not be totally re-intercalated into the host on subsequent discharge, and showed poor capacity retention as well. To gain a fundamental understanding for further improvement, neutron diffraction and electron microscopy were applied to investigate the causes of this phenomenon [46]. It was found that the layered structure had been transformed to a highly disordered spinel after just a few cycles. In the first charge process, Li ions were initially removed by a two-phase mechanism involving the original monoclinic layered phase and a hexagonal phase. Then a single hexagonal phase was observed between 30% and 100% of the Li ions in the original materials being removed. Finally, a dramatic collapse of the interlayer spacing was observed at the very end of the first charge process. Subsequent cycling raised the amount of spinel phase in the material, indicated by an increased capacity at the potential region of 4 V (Figure 10b) [47]. However, the rate of this transformation could be significantly decreased if a non-stoichiometric layered $Li_xMn_yO_2$ was used. After 100 cycles, both neutron diffraction and NMR analysis indicated that only 25% of the spinel structure was formed [48] [49]. Additional Co or Ni doping in the $NaMnO_2$ precursor could also slow the transformation from layered structure to spinel structure. The doped $LiM_yMn_{1-y}O_2$ (M = Co or Ni) materials showed improved capacity, rate capability, and cycling stability when compared with non-substituted $LiMnO_2$ [50] [51]. Although these results look promising, the preparation process presents problems for large-scale manufacturing.

An alternative approach to synthesizing layered m-$LiMnO_2$ at low temperatures is the hydrothermal method using permanganates and a lithium source as precursors [52]. The resulting $Li_xMnO_2{\cdot}nH_2O$ is dehydrated under mild conditions to give the desired layered Li_xMnO_2. The analogs doped by cobalt, iron, or nickel, Li_xMn_{1-y}(Co, Fe, or Ni)$_yO_2$, could also be synthesized by this method to obtain the same layered structure [53]. $Li_xMn_{0.99}Co_{0.01}O_2$ has shown a capacity between 0.7 and 0.8 Li/Mn, with good reversibility at a rate of 0.1 mA cm^{-2} for charging/discharging. However, cycling at high current densities, such as 1 mA cm^{-2}, could initiate a conversion from the layered phase to the spinel-like phase. Layered $LiMnO_2$ materials prepared at low temperatures often show much better capacities and rate capabilities when compared to those prepared at high temperatures. This is due to the shorter solid diffusion lengths for Li ions within these materials, which have lower crystallinity, smaller particle sizes, and higher surface areas. However, these materials often have low volumetric densities and therefore low energy densities. Furthermore, large surface areas have the potential to accelerate side reactions in the cathode/electrolyte interface, thereby lowering both safety and cycling stability.

Layered m-$LiMnO_2$ can also be prepared at high temperatures with an appropriate doping process. This kind of approach was first reported in 1995 by Davidson et al. [54]. Doped with trivalent metal ions, such as Al^{3+}, Ga^{3+}, and Cr^{3+}, $LiMnO_2$ could crystallize as a layered monoclinic phase (space group C2/m) by a simple solid-state reaction at 900-1000 °C under inert gas [47] [55] [56] [57] [58]. Although doping with Al or Ga improved the capacity retention of m-$LiMnO_2$, mainly through slowing the rate of crystal transformation from a layered to a spinel structure, complete prevention of this crystal transformation was not observed (Figure 10b). However, if the material was doped with 3% Cr, a large improvement in capacity retention was observed, as shown in Figure 10c [47]. A slight additional capacity can be achieved at 4 V even after 200 cycles, indicating almost no formation of spinel-type intercalation sites for Li ions. X-ray diffraction spectra taken from cycled Cr-substituted materials showed peaks for the hexagonal phase, with an unknown structure confirming the absence of the spinel phase. This might be interpreted as indicating that Cr^{3+}, with a strong stabilization energy for octahedral sites, could hinder the second stage of transformation from layer to spinel structures, as predicted by Ceder at el. [42].

In summary, layered m-$LiMnO_2$ materials with or without doping of other metal elements can be prepared at high temperatures using a solid-state reaction, or at low temperatures using an ion exchange or hydrothermal route.

However, most of the materials undergo structural transformation during battery cycling, forming defective spinel-type structures. Only Cr^{3+}-doping effectively inhibits the full transformation of the layered structure to the spinel structure, by hindering the second transformation stage.

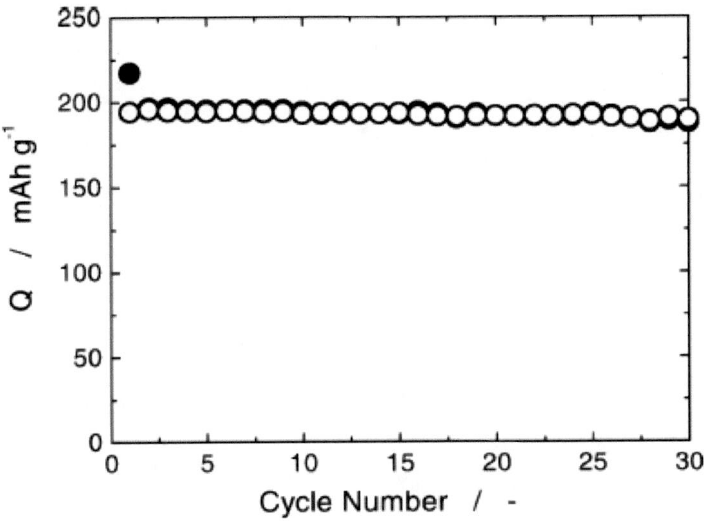

Figure 12. Charge (closed circles) and discharge (open circles) capacity fading upon cycling of Li/$LiNi_{0.5}Mn_{0.5}O_2$ cell operated between 2.5-4.5 V at a rate of 0.17 mA cm^{-2}. [73]

2.3.3 Development of orthorhombic $LiMnO_2$ cathode materials

In the structure of thermodynamically stable *o*-$LiMnO_2$ (space group Pmmn), illustrated in Figure 8b, Li and Mn ions are located in the octahedral sites in an alternating zig-zag configuration, with edge-sharing between [LiO_6] and [MnO_6] octahedra. *o*-$LiMnO_2$ materials with special electrochemical properties have been intensively investigated over the last decade. They can be prepared by several synthesis methods, using either low-temperature or high-temperature routes.

Low-temperature *o*-$LiMnO_2$ materials have been prepared by several synthesis methods. *o*-$LiMnO_2$ was first reported by Ohzuku et al. [59] with a relatively high capacity of 190 mA h g^{-1} over a voltage range of 2.0–4.25 V. This material was synthesized by heating mixed stoichiometric γ-MnOOH and LiOH at 300–450 °C under dry nitrogen. Reimers et al. [60] also reported a high-capacity *o*-$LiMnO_2$ synthesized by ion-exchanging of γ-MnOOH in boiling LiOH solution, with subsequent drying at 105 °C in air or heating at 200 °C under argon. However, the orthorhombic structure of the materials was found to gradually transform into a spinel structure. As a result, low-temperature *o*-$LiMnO_2$ materials showed poor capacity retention with extended cycling.

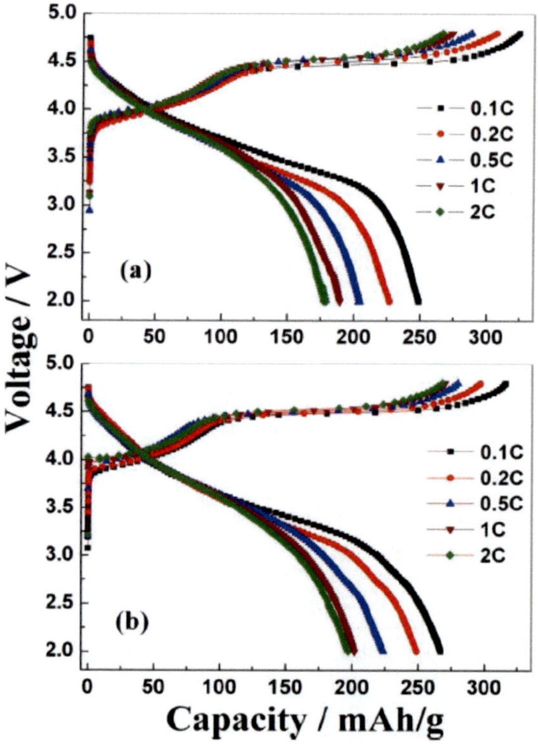

Figure 13. The voltage profiles upon capacity during the first charge/discharge process of the (a) pristine, and (b) AlF_3-coated $Li(Li_{0.2}Mn_{0.54}Ni_{0.13}Co_{0.13})O_2$ at different current density, 1C = 180 mA g^{-1}. Capacity fades fast with current density, which indicates the poor kinetic of $Li(Li_{0.2}Mn_{0.54}Ni_{0.13}Co_{0.13})O_2$ without or with AlF_3 coating. The ICL is successfully reduced from 75 to 47 mA h g^{-1} at 0.1C rate, and from 88 to 68 mA h g^{-1} at 2C rate. [91]

High-temperature *o*-$LiMnO_2$ materials can be synthesized by solid-state reaction of manganese oxide with lithium salt under an inert atmosphere. The high-temperature *o*-$LiMnO_2$ cathodes do have the problem of structural transformation during cycling [47], but generally give better capacity retention than low-temperature *o*-$LiMnO_2$. High-temperature *o*-$LiMnO_2$ showed an improved discharge capacity, from 160 to 200 mA h g^{-1}, when Mn_2O_3 instead of MnO_2 was used as the precursor [61] [62]. The crystallite size of the electrode materials also played an important role in cell performance. Detailed investigation into the effect of crystallinity on electrochemical performance confirmed that crystallite size was the critical parameter in determining the electrochemical performance of high-temperature *o*-$LiMnO_2$ phases [63] [64] [65]. Despite having good capacity retention, *o*-$LiMnO_2$ prepared at high temperatures also shows poor rate capability, as shown in Figure 11. Meanwhile, the sharp drop in discharge voltage, resulting from the thermodynamically stable spinel phase, is a problem in practical applications [39].

As described above, doping and coating of $LiMnO_2$-based cathode materials with orthorhombic or layered O_3 structures still does not prevent transformation into the spinel structure after extended cycling. This is partially due to the existence of the same ccp oxygen sub-lattice in all structures. If $LiMnO_2$-based cathode materials could be synthesized to yield a different oxygen sub-lattice, the transformation would require a coordinated lattice rearrangement. However, this rearrangement still could not be achieved at ambient temperature.

2.4 Mixed Transition Metal Dioxides

Another way to increase the doping level is to form mixed transition metal dioxides. These are another kind of cathode material with a layered structure. Ni and/or Co are used as substitutes for Mn to stabilize the structure, and simultaneously to increase the electronic conductivity of layered $LiMnO_2$ [66]. It has been found that the mixed transition metal dioxides tended to form $LiCoO_2$ (R-3m) analog with a layered structure, but not in solid solutions of $LiCoO_2$, $LiNiO_2$, and $LiMnO_2$. The valence states of the ions were Ni^{2+}, Co^{3+}, and Mn^{4+}. During charge-discharge cycling, the valence state of Mn^{4+} remained unchanged [67] [68].

Several representative mixed layered compounds will now be discussed:

(1) $LiNi_{0.5}Mn_{0.5}O_2$. Denoted as 550 material (0.5 Ni, 0.5 Mn, 0.0 Co), $LiNi_{0.5}Mn_{0.5}O_2$ was first reported by Rossen et al. [69] in 1992. Later, Spahr et al. [70] used X-ray photoelectron spectroscopy (XPS) and

magnetic data to determine that the nickel and manganese in the 550 material were in the forms of Ni^{2+} and Mn^{4+} ions rather than Ni^{3+} and Mn^{3+}. During electrochemical cycling, nickel was found to be the only active redox species cycling between the +2 and +4 valence states, while manganese remained in the +4 valence state, independent of lithium content. The stable Mn^{4+} could successfully prevent the Jahn-Teller effect coming from Mn^{3+} [71]. This was further confirmed by the charge-discharge curves, which showed a single-phase reaction similar to that of $LiNiO_2$ [72]. Layered $LiNi_{0.5}Mn_{0.5}O_2$ with optimum electrochemical performance was also synthesized by sintering nickel manganese double hydroxide precursors with lithium hydroxide at 1000 °C [73]. This yielded an initial discharge capacity of *ca.* 190 mA h g^{-1} between 2.5 V and 4.5 V at 0.17 mA cm^{-2}, and almost no deterioration in capacity was detected in the first 30 cycles (Figure 12) [74]. The rate capacity approached 190 mA h g^{-1} at 0.17 mA cm^{-2}, and declined to 135 mA h g^{-1} at 6 mA cm^{-2}. It is believed that pulse discharge rates over 10 mA cm^{-2} should be achievable according to the declining tendency indicated by the measurements [3]. Aluminum was also used as a doping metal [74]. Al doping on the Ni and Mn sites of the 550 materials reduced cation mixing and initial irreversibility, thus improving structural stability and capacity retention [75]. Moreover, Wang et al. [76] and Zhou et al. [77] proved that 10% Al-doped 550 materials had a better thermal stability than $LiCoO_2$, $LiMn_2O_4$, and $LiNi_1/3Mn_1/3Co_1/3O_2$, and a higher volumetric energy density than both $LiMn_2O_4$ and $LiFePO_4$ materials. Therefore, Al-doped $LiNi_{0.5-z}Mn_{0.5-z}Al_{2z}O_2$ may be candidates for cathode materials in large-size lithium ion cells for EV/HEV, where low cost, excellent safety, and high energy density are required.

(2) $LiNi_{1/3}Mn_1/3Co_1/3O_2$. Denoted as 333 material (0.33 Ni, 0.33 Mn, 0.33 Co), $LiNi_{1/3}Mn_1/3Co_1/3O_2$ was first reported by Liu et al. [78] in 1999. As discussed in section 2.2.3, the addition of cobalt to $LiMn_{1-y}Ni_yO_2$ can stabilize the layered structure. It was observed that using this cobalt-doped material, the volume change was less than 2% as the lithium ions were removed [79]. Many synthesis methods for 333 materials have been developed over a wide range of temperatures. Most of these materials showed similar electrochemical behavior; that is, the capacity could be increased by increasing the charge cut-off potential [80] [81]. Ohzuku et al. [82] [83] synthesized 333 material from

$LiOH \cdot H_2O$ and triple hydroxide precursors of cobalt, nickel, and manganese at 1000 °C, generating a material which could deliver a capacity of around 150 mA h g^{-1} when cycling between 2.5 and 4.2 V with a rate of 0.17 mA cm^{-2} at 30 °C. Increasing the charge cut-off potential to 4.6 V yielded a capacity in excessive of 200 mA h g^{-1}. 333 material prepared by the hydrothermal method showed a low capacity. However, after the material was sintered at 800 °C, enhanced capacities of 182 mA h g^{-1} at 0.2 C (30 mA/g), and 124 mA h g^{-1} at 5 C (750 mA/g) within the potential window of 2.8–4.6 V were achieved [[84]]. An increase in synthesis temperature from 800 to 900 °C also resulted in an improved initial capacity, from 173 to 190 mA h g^{-1} [[85]]. However, the 333 materials showed a large initial irreversible capacity (up to 20%) and then a marked capacity fading upon cycling. In addition, the rate capacities of these materials need to be improved before they will be viable for commercialization [79] [85].

(3) $LiM_{1-y}(Li_{1/3}Mn_{2/3})_yO_2$. $LiM_{1-y}(Li_{1/3}Mn_{2/3})_yO_2$ represents Li-rich layered compounds, where M can be Cr, Mn, Fe, Co, Ni, or mixtures thereof. These materials are solid solutions of layered $LiMO_2$ and $Li(Li_{1/3}Mn_{2/3})O_2$. The latter compound can also be treated as Li_2MnO_3, where Mn is Mn^{4+} rather than Mn^{3}; thus, any impact of the Jahn-Teller effect coming from Mn^{3+} would be minimized. This is why the stability of Li-rich 550 material, $Li_{1+x}(Ni_{0.5}Mn_{0.5})_{1-x}O_2$, can be increased by the addition of excess lithium [[86]].

(4) $yLiNiO_2 \bullet (1\text{-}y)Li[Li_{1/3}Mn_{2/3}]O_2$. This material can be treated as a solid solution of $LiNiO_2$-Li_2MnO_3. The structure and electrochemical performance of this material were systematically studied by Lu et al. [[87]]. The layered degree of this solid solution can be decreased by increasing the nickel content; the capacity decreases accordingly with decreasing y value (y = 1/3, 5/12, and 1/2) [87]. The capacity of the y = 1/3 material can be increased up to 200 mA h g^{-1} at 30 °C between 2.0 and 4.6 V. $Li(Li_{0.2}Ni_{0.2}Mn_{0.6})O_2$, described as $[LiNi_{0.5}Mn_{0.5}]_{0.4} \bullet [Li(Li_{1/3}Mn_{2/3})\ O_2]_{0.6}$, has also shown a steady-state capacity of around 200 mA h g^{-1} between 2.0 and 4.6 V at 0.1 mA cm^{-2} after the first 10 cycles [88]. This anomalously high capacity, far in excess of the theoretical value, can be obtained in $LiNi_{1/3}Co_{1/3}Mn_{1/3}O_2$ and Li_2MnO_3 solid solution [[88]]. $Li(Li_{0.2}Mn_{0.54}Ni_{0.13}Co_{0.13})O_2$, described as $[LiNi_{1/3}Co_{1/3}Mn_{1/3}O_2]_{0.4} \bullet [Li(Li_{1/3}Mn_{2/3})\ O_2]_{0.6}$, can deliver an initial

discharge capacity of 286 mA h g^{-1} between 4.8 V and 2.0 V at 0.05 mA cm^{-2} and 50 °C [[89]] [[90]]. Moreover, the addition of cobalt to the solid solution has been found to be helpful in retaining the rate capacity [[91]]. However, due to the oxidation of electrolytes at high potentials and attacks on the electrode/electrolyte interface from trace acidic species (HF) in the electrolyte, the pristine $Li(Li_{0.2}Mn_{0.54}Ni_{0.13}Co_{0.13})O_2$ cathode could suffer from a serious irreversible capacity loss (ICL) of 75 mA h g^{-1} at the 0.1 C rate (Figure 13), a poor rate capacity of 178 mA h g^{-1} at 2C (Figure 13), and a low capacity retention of 67.8% at the 0.5 C rate after 80 cycles (Figure 14) [[92]] [[93]]. Therefore, for commercialization of this material, the large ICL must be reduced and the low conductivity improved. Zheng et al. [91] [92] reported that AlF_3 or TiO_2 coating on the surface of lithium-rich compounds could suppress the side reaction between the electrolyte and the oxidative cathodes, and thus improve the electrochemical performance. After AlF_3 coating, the ICL was reduced to 47 mA h g^{-1} at the 0.1 C rate, and to 68 mA h g^{-1} from 88 mA h g^{-1} at the 2 C rate. Moreover, 87.9% of the initial specific discharge capacity was retained after 80 cycles at the 0.5 C rate at room temperature, 20.1% higher than that of pristine $Li(Li_{0.2}Mn_{0.54}Ni_{0.13}Co_{0.13})O_2$. Results obtained using in-situ differential electrochemical mass spectrometry revealed that the activity of oxygen species extracted from the $Li(Li_{0.2}Mn_{0.54}Ni_{0.13}Co_{0.13})O_2$ cathode was greatly reduced and the decomposition of the electrolyte was appreciably suppressed after AlF_3 coating, which could significantly improve the safety of Li cells when using this kind of cathode. For $yLiNiO_2 \cdot (1\text{-}y)Li[Li_{1/3}Mn_{2/3}]O_2$ to be a viable cathode material for batteries, its rate capability needs further improvement and its lifetime should be evaluated.

3 Spinel Lithium Manganese Oxides

3.1 Introduction

The $A[B_2]O_4$ spinel structure (Figure 15) has a ccp oxygen lattice closely related to the α-$NaFeO_2$ layer structure, differing only in the distribution of the cations among the available octahedral and tetrahedral sites [[94]]. The A cations occupy the 8a tetrahedral sites, and the B cations occupy the 16d octahedral sites. The $[B_2]O_4$ array forms a strongly bonded 3D framework in which the 8a tetrahedral sites and 16c octahedral sites create a 3D interconnected interstitial space for lithium transportation [95].

Two promising spinel compounds, $LiMn_2O_4$ and $Li_4Ti_5O_{12}$, have been developed as electrode materials for lithium ion batteries. $LiMn_2O_4$ shows a 4 V plateau during lithium intercalation/de-intercalation, making it a feasible cathode material. $Li_4Ti_5O_{12}$, due to its low lithium intercalation/de-intercalation voltage (1.5 V), may be more suitable as an anode material.

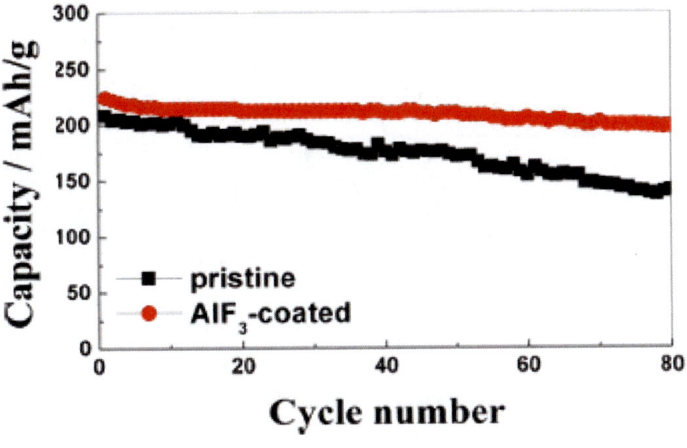

Figure 14. The specific discharge capacity upon cycling of the pristine (squares), and AlF_3-coated (circles) $Li(Li_{0.2}Mn_{0.54}Ni_{0.13}Co_{0.13})O_2$ at room temperature over the voltage range of 2.0-4.8V at room temperature and 0.5C rates, 1C = 180 mA g^{-1}. 67.8% of initial specific discharge capacity was retained upon 80 cycles at 0.5C rate for the pristine sample, and increased to 87.9% for AlF_3 coated sample. [92]

3.2 $LiMn_2O_4$

3.2.1 Problems with $LiMn_2O_4$

$LiMn_2O_4$ has a spinel structure with cubic symmetry and a space group of Fd-3m (No. 227), locating Li at the 8a tetrahedral sites and Mn at the 16d octahedral sites. The 16c octahedral sites are open and face-shared with 8a tetrahedral sites, affording fast lithium transportation along the 8a-16c-8a path. The cubic symmetric $[Mn_2]O_4$ framework can undergo isotropic expansion and shrinkage upon lithium intercalation/de-intercalation, resulting in a high intercalating/de-intercalating reversibility [95] [96]. The theoretical specific capacity of $LiMn_2O_4$ is 148 mA h g^{-1}. However, in practice, only 0.85 Li/Mn can be removed electrochemically between 3.5 V and 4.3 V, delivering a specific capacity of 130 mA h g^{-1} at 4.0 and 4.15 V (Figure 16) [97]. A second lithium intercalation/de-intercalation in the 16a sites of $LiMn_2O_4$ will produce a 3 V plateau similar to that of layered $LiMnO_2$ after a long cycling, as discussed in section 2.3.1. However, a second lithium intercalation could cause severe Jahn-Teller distortion and transformation to a tetragonal phase. Usually only the 4 V plateau of $LiMn_2O_4$ is used in lithium ion batteries.

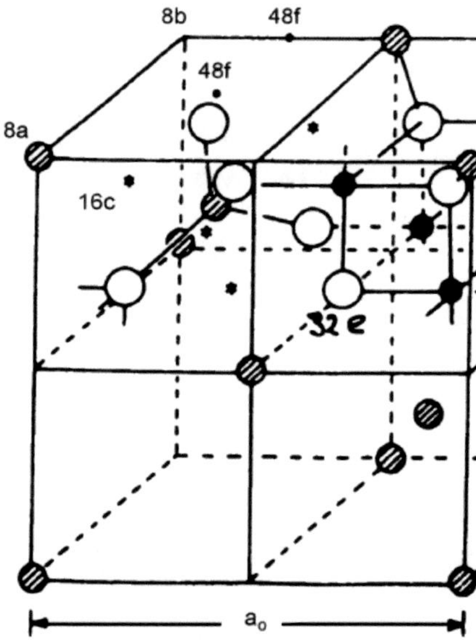

Figure 15. Two quadrants of the $A[B_2]O_4$ spinel structure, with A (Striate balls) at 8a sites, B (Black alls) at 16d sites, and O (White balls) at 32e sites. For $LiMn_2O_4$, Li will occupy 8a sites and Mn at 16d sites. [93]

First reported by Thackeray et al. [98], $LiMn_2O_4$ is cost-effective, environmentally friendly, and thermally stable, features which make it a promising cathode material for high-power lithium ion batteries in HEV/EV applications. Unfortunately, the rate capacity and capacity retention, particularly at elevated temperatures, are insufficient and need to be enhanced prior to commercial use. It has been found that the capacity fading is mainly caused by (1) dissolution of Mn^{3+} by disproportional reaction with Mn^{4+} and electrolyte-soluble Mn^{2+}, (2) phase transformation from cubic to tetragonal symmetry by Jahn-Teller distortion of [MnO_6], and (3) the high oxidation ability of Mn^{4+} on organic electrolyte [95] [[99]]. Furthermore, in comparison with the rate capability of $LiCoO_2$, that of $LiMn_2O_4$ is inferior due to its low electric conductivity (10^{-6} S/cm).

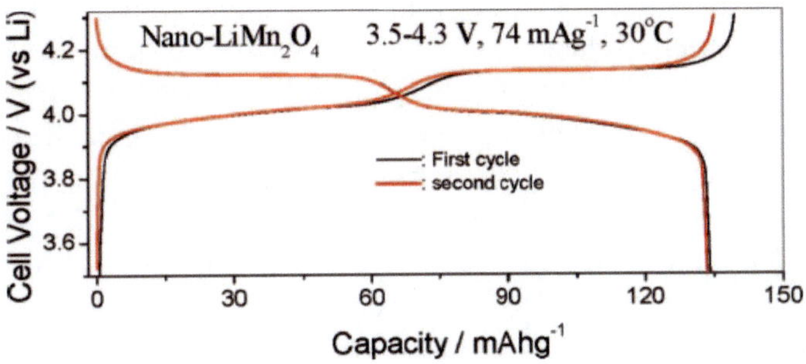

Figure 16. Charge and discharge voltage profiles for nano-$LiMn_2O_4$ at 0.5C rate between 3.5-4.3 V for the first and second cycles at 30 °C. [96]

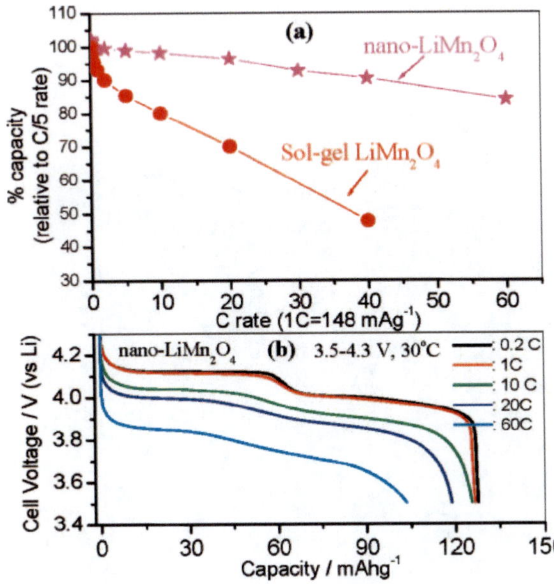

Figure 17. Rate performance of nano-$LiMn_2O_4$ compared with sol-gel $LiMn_2O_4$. The rate capability is expressed as the percentage of the capacity obtained at a specific discharge rate compared to that obtained at 0.2C rate (30 mA g^{-1}; around 0.15 mA cm^{-2}). (b) The discharge voltage profile for nano-$LiMn_2O_4$ at different C-rates. [96]

3.2.2 Modification of $LiMn_2O_4$

The insufficient cycle life of $LiMn_2O_4$-based cathodes is believed to be caused by the dissolution of divalent manganese ions formed by the disproportional reaction of trivalent manganese. Therefore, it is very important to minimize the amount of trivalent manganese formed during cycling. Tarascon et al. [100] found that the value of the cubic lattice parameter (a_0) in $LiMn_2O_4$ was directly related to the average valence state of the manganese, which was critical for obtaining high performance. When the oxidation state of the manganese was 3.58 or higher, such as in lithium-rich $Li_{1+x}Mn_{2-x}O_4$ with a_0 = 8.23 Å or less, the dissolution of manganese could be minimized and the effect of the Jahn-Teller distortion associated with the Mn^{3+} ion was reduced as well [101] [102]. On the other hand, storage performance also needs to be seriously considered for lithium batteries using spinel $LiMn_2O_4$ as the cathode material. In addition, $LiMn_2O_4$ also suffers from severe self-discharge in its fully charged state, particularly at elevated temperatures, due to the redox reaction between tetravalent manganese and organic electrolyte solvent [103] [104]. It has also been determined that the presence of trace moisture in fluoride-containing electrolyte can generate HF, which then dissolves the spinel oxides. This problem can be solved using electrolyte containing a non-fluoride lithium salt, such as LiBOB, rather than $LiPF_6$ [105].

Figure 18. Crystal structure of olivine $LiMPO_4$,where M is either Mn, Fe, Co, or Ni, constructed by [MO_6] (octahedra), [PO_4] (tetrahedral), and the Li atoms (ball). [116]

Figure 19. SEM images of (A) an Fe(II) phosphate and lithium phosphate, co-precipitated from aqueous solutions, and of (B) lithium iron phosphate, synthesized thereof by heat treatment in nitrogen. [127]

Co-doping is an effective way to improve the performance of $LiMn_2O_4$-based cathode materials. For example, co-doping with aluminum and fluoride in lithium-rich $Li_{1+x}Mn_{1-x-y}Al_yO_{4-z}F_z$ can enhance the capacity retention at elevated temperatures [106]. Hundreds of studies have been published on $LiMn_2O_4$ treated with various kinds of dopants to stabilize its structure, order its cation distribution, reduce the proportion of trivalent manganese, and increase its conductivity [3].

Another solution is to coat the $LiMn_2O_4$ particle surface with a conductive or protective material such as Ag [107], MgO [108], Al_2O_3 [109], ZnO [110], ZrO_2 [111], or $Li_{1-x}CoO_2$ ($x \geqslant 0$) [112] [113] [114]. Li_xCoO_2 ($0 < x \leqslant 1$) coating can be applied to the surface of $LiMn_2O_4$ by sol-gel methods [112] or a micro-emulsion method [113]. Further heat treatment at 800 °C can lead to the formation of a core-shell structure with a new spinel phase, $Li_{1+x}Mn_{2-x}Co_xO_4$, on the surface [112]. In this $Li_{1+x}Mn_{2-x}Co_xO_4$ coating, cobalt tends to be divalent, which can reduce the amount of trivalent manganese and therefore prevent manganese from dissolving into the electrolyte solution. In addition, the $Li_{1+x}Mn_{2-x}Co_xO_4$ coating layer also exhibited a lower resistance than the Co-doped spinel material, which was confirmed by electrochemical impedance spectroscopy (EIS). As a result, the cycling capacity and rate capacity of coated $LiMn_2O_4$ were greatly improved at both room and elevated temperatures. The capacity retention at a 20 C rate could be increased from 50% to 80% at the maximum capacity.

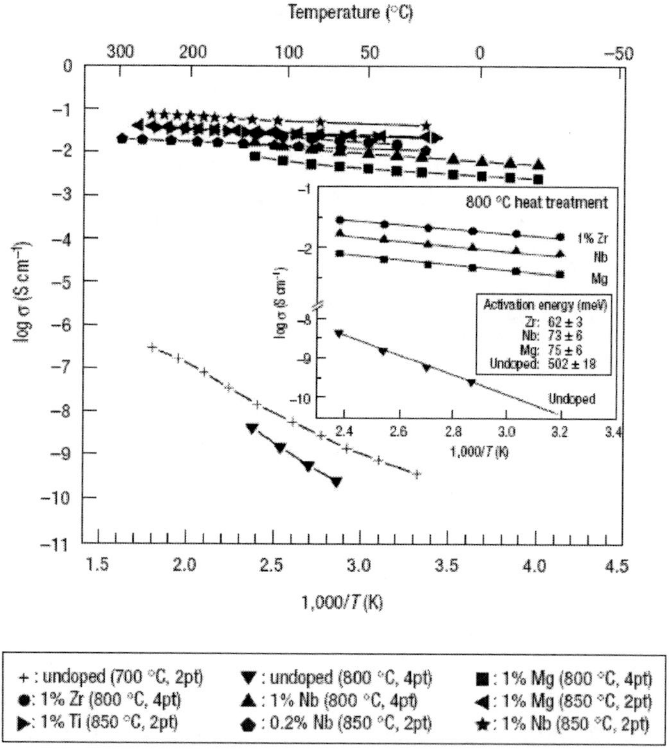

Figure 20. Doped olivines of stoichiometry $Li_{1-x}M_xFePO_4$ show electrical conductivity at room temperature that is a factor of ca. 10^8 greater than in undoped $LiFePO_4$, and absolute values > 10^{-3} S cm^{-1} over the temperature range -20 °C to +150 °C of interest for battery applications. Results are for polycrystals fired at 700-850 °C and measured by two-point d.c. and four-point van der Pauw methods. Inset shows expanded plot for series of dense, single-phase samples fired at 800 °C, showing lower activation energy of the doped compositions. [122]

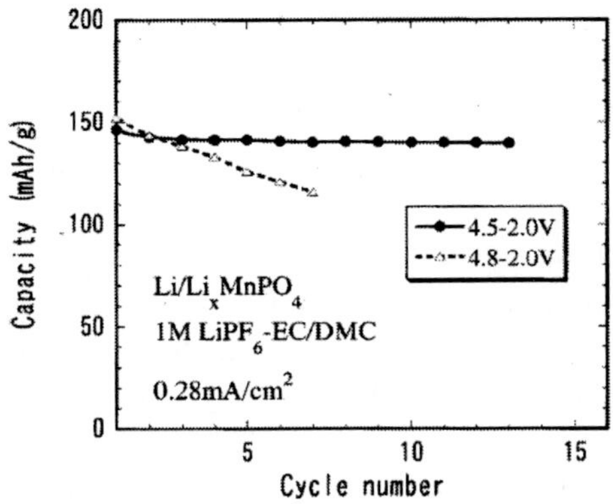

Figure 21. Cycling performance of the Li/C-Li_xMnPO_4 cell at 0.28 mA cm^{-2} at room temperature. [144]

Recently, Shaju et al. [97] reported a stoichiometric nano-$LiMn_2O_4$ prepared by the sol-gel method using RF resin, followed by calcination at 750 °C for 15 h. As shown in Figure 17, the material exhibits a high initial capacity (131 mA h g^{-1}) and retains 118 mA h g^{-1} after 200 cycles at a 0.5 C rate. It also exhibits an excellent rate capability (retaining 90% of its capacity at 40 C and 85% at 60 C), a nearly 100% power retention rate (5840 W kg^{-1} dropping to 5828 W kg^{-1} at 10 C after 1000 cycles), and a high volumetric energy density (around 750 Wh L^{-1} at 10 C). The excellent performance of nano-$LiMn_2O_4$ may be due to its stable carbonized surface that inhibits dissolution.

Spinel $LiMn_2O_4$ is therefore one of the most promising cathode candidates for high-power lithium batteries in terms of cost, toxicity, and safety. However, considering its low gravimetric energy density, spinel $LiMn_2O_4$ may not be suitable as a cathode material in portable batteries.

4 Olivine Lithium Metal Phosphates

4.1 Introduction

Padhi et al. [115] [116] [117] first reported the polyanionic compounds with an olivine or NASICON structure as insertion materials for lithium ion batteries in 1996. Olivine $LiFePO_4$ and NASICON $Li_3Fe_2(PO_4)_3$ are two typical compounds in this category [118].

Olivine lithium metal phosphates ($LiMPO_4$, where M is either Mn, Fe, Co, or Ni) have an orthorhombic symmetry with the space group *Pmna* (No. 52), as shown in Figure 18 [116]. The olivine structure consists of a slightly distorted hexagonal close-packed (hcp) framework, with Li and M in the octahedral 4a and 4c sites, and P in the tetrahedral 4c sites, respectively. The octahedral 4a sites form linear chains of edge-sharing [LiO_6] octahedra along the *b* axis, which can produce a 1D channel for lithium ion transportation. The octahedral 4c sites form staggered lines of corner-sharing [MO_6] octahedra along the *b* axis. The strong P-O covalency can stabilize the anti-bonding M^{3+}/M^{2+} state through the M-O-P inductive effect, generating a higher redox potential. Moreover, the MPO_4 3D framework with strong P-O covalency is stable not only at high temperatures, but also during lithium intercalation/de-intercalation cycling.

Olivine $LiMPO_4$ sustains its orthorhombic structure after removal of one lithium ion, exhibiting only a 6.81% shrinkage in volume, which means it offers a good cycle life and excellent safety when used as the cathode in lithium ion batteries. However, separation of the [MO_6] octahedra by [PO_4] tetrashedra can dramatically reduce the material's electronic conductivity, leading to a poor rate capacity [119]. Great efforts have thus been made to improve the conductivity and hence the capacity of olivine phosphates. These cathode materials include $LiFePO_4$, $LiMnPO_4$, and $LiCoPO_4$. The electronic conductivity of $LiMPO_4$ olivines could be effectively improved by various techniques, such as carbon coating, cation doping, and synthesis of nanocrystalline grains [120] [121] [122] [123] [124] [125].

4.2 $LiFePO_4$

4.2.1 Problems with $LiFePO_4$

Phospho-olivines as lithium intercalation materials were reported early in 1997. Unfortunately, researchers did not pay much attention to them because olivine phosphates with low electronic conductivity did not allow most of the lithium ions to intercalate/de-intercalate reversibly. This low reversibility could lead to low capacity, particularly at high current densities. Under the strong driving force of safety requirements when lithium ion batteries are used commercially for both portable devices and EV/HEV applications, olivine phosphate cathode materials, especially $LiFePO_4$, have been revisited globally by researchers since the beginning of the 21st century. Attractive features are their low cost, low toxicity, excellent thermal stability, and promising electrochemical performance.

Olivine $LiFePO_4$ has an orthorhombic symmetry with the space group *Pmna* and cell parameters a = 10.333 Å, b = 6.011 Å, and c = 4.696 Å [126]. It can reversibly intercalate/de-intercalate almost all of the lithium ions at around 3.4 V vs. Li/Li^+, delivering a theoretical specific capacity 170 mA h g^{-1} and a theoretical energy density of 550Wh/kg. Due to the inductive effect from strong P-O covalency, olivine phosphates can maintain their structure during the lithium intercalation/de-intercalation process in coexistence with $LiFePO_4$ and $FePO_4$ [127]. Delithiated $FePO_4$ shows an olivine structure with a very slight deviation in cell parameters from that of $LiFePO_4$, which guarantees excellent stability of the crystal structure and thus excellent cycling performance [128]. The reversible capacity of $LiFePO_4$ can reach 170 mA h g^{-1}, with no obvious fading apparent even after several hundred cycles. Moreover, $LiFePO_4$ shows high thermal stability and low oxidative ability with the electrolyte [129] [130]. Unfortunately, the electronic conductivity of pure $LiFePO_4$ (ca. 10^{-9} S cm^{-1}) at room temperature [123] [125] was found to be far lower than that of $LiCoO_2$ (ca. 10^{-3} S cm^{-1}) [131] and of $LiMn_2O_4$ (ca. 10^{-4} S cm^{-1}) [132]. The poor kinetics of lithium intercalation in $LiFePO_4$, caused by a low lithium diffusion coefficient [133] and low electronic conductivity [134], could also restrict its practical capacity at high current densities, particularly limiting its use in high-power lithium ion batteries for EV/HEV applications. The other problem with $LiFePO_4$ is the difficulty of preventing Fe^{2+} from being oxidized to Fe^{3+} during preparation. Many studies of $LiFePO_4$ have thus been focused on preparation methods and on enhancing its electrochemical performance through doping and/or coating.

4.2.2 Synthesis methods for $LiFePO_4$

$LiFePO_4$ can be prepared by high-temperature solid-state reactions, or by low-temperature liquid methods. The first reported $LiFePO_4$ was synthesized by a high-temperature reaction of solid-state precursors, including ferrous salt, in an inert atmosphere [113] [125] [135]. Solid-state reactions can readily produce aggregative particles and some impurities due to incomplete mixing of reactants as well as lithium evaporation at high temperatures. The intimate grinding of starting materials can be attained by the mechanochemical activation method, resulting in pure $LiFePO_4$ with smaller particle sizes [134]. $LiFePO_4$ with small particle sizes can also be prepared at a lower temperature and in a shorter time period by a microwave method [136]. In addition, a novel carbothermal reduction method was also developed to synthesize $LiFePO_4$

using a low-cost ferric salt [137]. It was found that carbon resources in the starting materials not only acted as reductive reagents for the transition of Fe^{3+} to Fe^{2+} at high temperatures, but also prevented the final products from aggregating. Furthermore, the residual carbon film on the surface of $LiFePO_4$ could significantly increase its electronic conductivity.

Normally, low-temperature liquid methods consume less energy than high-temperature solid-state methods. $LiFePO_4$ could be synthesized by many low-temperature methods, such as the hydrothermal [123] [138] [139], co-precipitation [127] [134], or sol-gel method [134] [140]. Although the hydrothermal method is easy to operate, the synthesized olivine $LiFePO_4$ has shown poor electrochemical properties, as about 7% of the iron atoms occupied the lithium sites and blocked the diffusion of lithium ions [138]. Heat treatment of hydrothermal material at 700 °C ordered the lithium and iron atoms. Using the co-precipitation method, a finely dispersed precursor containing a stoichiometric composite of Fe(II) phosphate and lithium phosphate can be obtained. Sintering of this precursor can produce finely dispersed, rhombus shaped $LiFePO_4$ sheets (Figure 19). This material, when added with 20 wt% conductive carbon and 10 wt% PTFE, can deliver a constant capacity of 150 mA h g^{-1} at a 0.05 C rate in the first 40 cycles [127]. However, the amount of additives in the electrode composite needs to be reduced to further improve the specific capacity.

4.2.3 Electrochemical performance upgrading of $LiFePO_4$

As pure $LiFePO_4$ has a very low electronic conductivity (ca. 10^{-9} S cm^{-1}) at room temperature [122], the theoretical capacity (170 mA h g^{-1}) can only be observed at either a very low current density [120] or elevated temperatures [128]. Two strategies, carbon coating and cation doping, are generally adopted to improve the electronic conductivity and thereby the electrochemical performance of $LiFePO_4$.

Huang et al. [141] reported a $LiFePO_4$/C (15 wt% of carbon) composite and obtained capacities of 162 mA h g^{-1} at a 0.1 C rate and 110 mA h g^{-1} after 800 cycles at a 5 C rate, even with a low cathode loading (5 mg/cm^2) and rather high carbon content (20 wt%, including 5 wt% additive carbon). Such a large amount of carbon could significantly decrease the tap density and energy density of $LiFePO_4$/C composites. Chen at al. [142] proposed a $LiFePO_4$/C (3.5 wt% of carbon) composite using sugar as a carbon precursor, achieving a rate capability comparable to that of $LiFePO_4$/C (15 wt%) reported by Huang et al. [140].

Chuang et al. [122] at MIT reported an impressive electrochemical performance for $LiFePO_4$ doped by cations, including Mg^{2+} and Nb^{5+}. The doped material showed an electronic conductivity increased by 8 orders of magnitude (Figure 20). At low rates (C/10 to C/30), this material gives a capacity of ~150 mA h g^{-1}, corresponding to ~90% of the theoretical value. The capacity is reduced to 100 mA h g^{-1} at 4.3 C, in the voltage range of 2.8–4.2 V at room temperature. Power and energy densities for a complete cell could be estimated as 1,300-2,200 W kg^{-1} and 32-53 W h kg^{-1} at a 20 C rate, and 2,800-4,670 W kg^{-1} and 18-30 Wh kg^{-1} at a 40 C rate, respectively. This result makes these low-cost and ultra-safe olivine materials very attractive for EV/HEV applications.

4.3 $LiMPO_4$ (M = Mn, Co, Ni)

Besides iron, the transition metal in olivine $LiMPO_4$ may also be manganese, nickel, or cobalt. However, no reports have been published on these materials that indicate an electrochemical performance superior to that of $LiFePO_4$, even though the materials have higher discharge potentials, as shown by experimental [143] [144] [145] and theoretical results [146]. The open-circuit voltages are calculated as 3.5 V for $LiFePO_4$, 4.1 V for $LiMnPO_4$, 4.8 V for $LiCoPO_4$, and 5.1 V for $LiNiPO_4$. Because no appropriate organic electrolyte can be sustained at a potential over 5 V, $LiNiPO_4$ is still not suitable to use as a cathode material for lithium ion batteries, at this current stage in our technology.

Olivine $LiCoPO_4$ can give a high lithium intercalation voltage plateau at 4.8 V with a capacity of 100 mA h g^{-1} at 0.2 mA cm^{-2} [142], which is about 60% of the theoretical capacity (167 mA h g^{-1}). Unfortunately, cobalt salts are costly, resulting in $LiCoPO_4$ being expensive as well and making it a less than ideal cathode candidate for practical applications despite its high lithium intercalation plateau. With respect to cost, spinel $LiNi_{0.5}Mn_{1.5}O_4$ material, which also has a lithium intercalation plateau of 4.8 V, may be more interesting than $LiCoPO_4$ [147].

Low-cost $LiMnPO_4$, with a lithium intercalation plateau of 4.1 V, is an attractive cathode candidate for lithium ion batteries. An earlier report on $LiMnPO_4$ indicated a very low capacity, which might be attributed to its low electronic conductivity or lithium ion conductivity [148]. $LiMnPO_4$ synthesized by both the direct precipitation method [149] and the hydrothermal method [150] did not show a favorable capacity, either. In this kind of material, Mn^{2+} is disordered on the Li^+ sites, which hinders lithium diffusion within the structure, similar to the case of hydrothermally prepared $LiFePO_4$. Carbon coating, reported by Li et al. [145], could improve the capacity of olivine $LiMnPO_4$ up to about 140 mA h g^{-1} with a current density of 0.28 mA cm^{-2} at room temperature (Figure 21). However, carbon-coated $LiMnPO_4$ showed a poor rate capacity of about 50 mA h g^{-1} at a 2 C rate [[151]], which was far lower than that of carbon-coated $LFePO_4$. Therefore, it is necessary to further improve the performance of manganese olivine when it is used as a cathode material for lithium ion batteries, even though it has a high lithium intercalation voltage and needs mild synthetic conditions.

5 Conclusion

Lithium ion battery technology still faces some challenges, including safety issues and insufficient energy density for modern electronic devices, although these batteries have been commercially available for many years. The emerging market of electric vehicles, hybrid electric vehicles, and plug-in hybrid electric vehicles (PHEV) is a strong driving force to overcome these challenges and stimulate the development of a new generation of lithium ion batteries with high energy density and high power density. In order to address these issues, cathode materials in lithium ion batteries are the primary targets of research and development. In previous decades, layered $LiCoO_2$ was the cathode material for the first generation of lithium ion batteries, and even today it is still used in most of commercially available lithium ion batteries. But because of its poor thermal stability and low specific capacity, alternative new inorganic cathode materials are urgently needed to replace $LiCoO_2$. With rapid research and development, several kinds of materials have been explored, including $LiNiO_2$, $LiMnO_2$, $LiMn_2O_4$, and $LiFePO_4$.

Compared with $LiCoO_2$, layered $LiNiO_2$-based compounds show higher specific capacity but poorer thermal stability. Although lattice doping and surface coating can further improve performance, these materials can only be used in limited fields and are not suitable for completely replacing $LiCoO_2$ in the main lithium ion battery market. Layered $LiMnO_2$-based compounds are not considered practical cathode materials because of their structural instability. Fortunately, layered mixed transition metal oxides, particularly the 550 and 333 materials, display higher specific capacity and better thermal stability than $LiCoO_2$, and should be able to gradually replace $LiCoO_2$ in lithium ion batteries for portable electronic devices.

Compared with $LiCoO_2$, spinel $LiMn_2O_4$ is another promising cathode material, being low-cost, environmentally benign, and thermally stable. Unfortunately, its low rate capability and poor capacity retention at elevated temperatures compromise this material's suitability. Efforts to improve the electrochemical performance of spinel $LiMn_2O_4$ through material modifications have recently intensified.

Olivine $LiFePO_4$ is another good cathode material candidate because of its low cost, low toxicity, and excellent thermal stability, its major drawback being low electronic conductivity. With respect to this, great progress has been made using lattice doping and surface coating technologies. Because of its affordability and safety, $LiFePO_4$ has emerged as a commercially available new cathode material for the next generation of lithium ion batteries, particuraly in EV/HEV/PHEV applications.

REFERENCES

[1] Tarascon, JM; Armand, M. *Nature.,* 2001, 414, 359-367.
[2] Whittingham, MS. *Chem. Rev.*, 2004, 104, 4271-4301.
[3] Mitzushima, K; Jones, PC; Wiseman, PJ; Goodenough, JB. *Mater. Res. Bull.*, 1980, 15, 783-789.
[4] Ozawa, K. *Solid State Ionics,* 1994, 69, 212-221.
[5] Amatucci, GG; Tarascon, JM; Klein, LC. *J. Electrochem. Soc.*, 1996, 143, 1114-1123.
[6] Kim, J; Noh, M; Cho, J; Kim, HM; Kim, KB. *J. Electrochem. Soc.*, 2005, 152, A1142-A1148.
[7] Li, G; Yang, ZX; Yang, WS. *J. Power Sources.,* 2008, 183, 741-748.
[8] Gabrisch, H; Yazimi, R; Fultz, B. *J. Electrochem. Soc.*, 2004, 151, A891-A897.
[9] Shao-Horn, Y; Hackney, SA; Armstrong, AR; Bruce, PG; Gitzendanner, R; Johnson, CS; Thackeray, MM. *J. Electrochem. Soc.*, 1999, 146, 2404-2412.
[10] Hwang, SJ; Park, HS; Choy, JH; Campet, G; Portier, J; Kwon, CW; Etourneau, J. *Electrochem. Solid-State Lett.*, 2001, 4, A213-A216.
[11] Armstrong, AR; Robertson, AD; Dupre, N; Grey, CP; Bruce, PG. *Chem. Mater.*, 2004, 16, 3106-3118.

[12] Patoux, S; Dolle, M; Doeff, MM. *Chem. Mater.*, 2005, 17, 1044-1054.
[13] Broussely, M; Perton, F; Biensan, P; Bodet, JM; Labat, J; Lecerf, A; Delmas, C; Rougier, A; Pérès, JP. *J. Power Sources.,* 1995, 54, 109-114.
[14] Lu, CH; Cheng, L. *J. Mater. Chem.*, 2000, 10, 1403-1407.
[15] Barboux, P; Tarascon, JM; Shokoohi, FK. *J. Solid State Chem.*, 1991, 94, 185-196.
[16] Schoonman, J; Tuller, HL; Kelder, EM. *J. Power Sources.,* 1999, 81, 44-48.
[17] Peres, JP; Delmas, C; Rougier, A; Broussely, M; Perton, F; Biensan, P; Willmann, P. *Phys. Chem. Solids.,* 1996, 57, 1057-1060.
[18] Ohzuku, T; Komori, H; Swai, K; Hirai, T. *Chem. Express.,* 1990, 5, 733-736.
[19] Ohzuku, T; Veda, A; Nagayama, M. *J. Electrochem. Soc.*, 1996, 140, 1862-1869.
[20] Delmas, C; Menetrier, M; Croguennec, L; Levasseur, S; Peres, JP; Pouillerie, C; Prado, G; Fournes, L; Weill, F. *International Journal of Inorganic Materials.,* 1999, 1, 11-19.
[21] Peres, JP; Weill, F; Delmas, C. *Solid State Ionics.,* 1999, 116, 19-27.
[22] Imhof, R; Novak, P. *J. Electrochem. Soc.*, 1999, 146, 1702-1706.
[23] Saadoune, I; Delmas, C. *J. Solid State Chem.*, 1998, 136, 8-15.
[24] Saadoune, I; Menetrier, M; Delmas, C. *J. Mater. Chem.*, 1997, 7, 2505-2511.
[25] Nakai, I; Nakagome, T. *Electrochem. Solid-State Lett.*, 1998, 1, 259-261.
[26] Liu, H; Li, J; Zhang, Z; Gong, Z; Yang, Y. *Electrochim. Acta.,* 2004, 49, 1151-1159.
[27] Liu, H; Li, J; Zhang, Z; Gong, Z; Yang, Y. *J. Solid State Electrochem.*, 2003, 7, 456-462.
[28] Fey, GTK; Chen, JG; Subramanian, V. *J. Power Sources.,* 2003, 119, 658-663.
[29] Broussely, M; Biensan, P; Simon, B. *Electrochim. Acta.,* 1999, 45, 3-22.
[30] Arai, H; Tsuda, M; Saito, K; Hayashi, M; Takei, K; Sakurai, Y. *J. Solid State Chem.*, 2002, 163, 340-349.
[31] Guilmard, M; Croguennec, L; Denux, D; Delmas, C. *Chem. Mater.*, 2003, 15, 4476-4483.
[32] Guilmard, M; Croguennec, L; Delmas, C. *Chem. Mater.*, 2003, 15, 4484-4493.
[33] Zhang, ZR; Liu, HS; Gong, ZL; Yang, Y. *J. Electrochem. Soc.*, 2004,

151, A599-A603.
[34] Kweon, HJ; Kim, SJ; Park, DG. *J. Power Sources*., 2000, 88, 255-261.

[35] Lee, HY; Hong, YS; Kim, YM; Kim, G; Shin, NS; Cho, J. *J. Electrochem. Soc*., 2006, 153, A781-A786.
[36] Cho, J; Kim, TJ; Kim, J; Noh, M; Park, B. *J. Electrochem. Soc*., 2004, 151, A1899-A1904.
[37] Liu, H; Zhang, Z; Gong, Z; Yang, Y. *Solid State Ionics.,* 2004, 166, 317-325.
[38] Ammundsen, B; Paulsen, J. *Adv. Mater*., 2001, 13, 943-956.
[39] Ceder, G; Van der Ven, A. *Electrochim. Acta.,* 1999, 45, 131-150.
[40] Kotschau, IM; Dahn, JR. *J. Electrochem. Soc*., 1998, 145, 2672-2677.
[41] Ceder, G; Mishra, SK. *Electrochem. Solid-State Lett*., 1999, 2, 550-552.
[42] Fuchs, B; Kemmler-Sack, S. *Solid State Ionics.,* 1994, 68, 279-285.
[43] Armstrong, AR; Bruce, PG. *Nature.,* 1996, 381, 499-500.
[44] Capitaine, F; Gravereau, P; Delmas, C. *Solid State Ionics.,* 1996, 89, 197-202.
[45] Shao-Horn, YS; Hackney, A; Armstrong, AR; Bruce, PG; Gitzendanner, R; Johnson, CS; Thackeray, MM. *J. Electrochem. Soc*., 1999, 146, 2404-2412.
[46] Ammundsen, B; Desilvestro, J; Groutso, T; Hassell, D; Metson, JB; Regan, E; Steiner, R; Pickering, PJ. *J. Electrochem. Soc*., 2000, 147, 4078-4082.
[47] Armstrong, AR; Dupre, N; Paterson, AJ; Grey, CP; Bruce, PG. *Chem. Mater*., 2004, 16, 3106-3118.
[48] Armstrong, AR; Paterson, AJ; Dupre, N; Grey, CP; Bruce, PG. *Chem. Mater*., 2007, 19, 1016-1023.
[49] Armstrong, AR; Gitzendanner, R; Robertson, AD; Bruce, PG. *Chem. Commun*., 1998, 1833-1834.
[50] Quine, TE; Duncan, MJ; Armstrong, AR; Robertson, AD; Bruce, PG. *J. Mater. Chem*., 2000, 10, 2838-2841.
[51] Chen, R; Whittingham, MS. *J. Electrochem. Soc*., 1997, 144, L64-L67.
[52] Zhang, F; Whittingham, MS. *Electrochem. Solid-State Lett*., 2000, 3, 309-311.
[53] Davidson, IJ; McMillan, RS; Murray, JJ. *J. Power Sources.,* 1995, 54, 205-208.
[54] Davidson, IJ; McMillan, RS; Slegr, H; Luan, B; Kargina, I; Murray, JJ;

Swainson, IP. *J. Power Sources.,* 1999, 81-82, 406-411.
[55] Jang, YI; Huang, B; Chiang, YM; Sadoway, DR. *Electrochem. Solid-State Lett*., 1998, 1, 13-16.
[56] Wang, H; Jang, YI; Chiang, YM. *Electrochem. Solid-State Lett*., 1999, 2, 490-493.
[57] Chiang, YM; Sadoway, DR; Jang, YI; Huang, B; Wang, H. *Electrochem. Solid-State Lett*., 1999, 2, 107-110.
[58] Ohzuku, T; Ueda, A; Hirai, T. *Chem. Express.,* 1992, 7, 193-196.
[59] Reimers, JN; Fuller, EW; Rossen, E; Dahn, JR. *J. Electrochem. Soc*., 1993, 140, 3396-3041.
[60] Croguennec, L; Deniard, P; Brec, R. *J. Electrochem. Soc*., 1997, 144, 3323-3330.
[61] Gummow, RJ; Liles, DC; Thackeray, MM. *Mater. Res. Bull*., 1993, 28, 1249-1256.
[62] Croguennec, L; Deniard, P; Brec, R; Lecerf, A. *J. Mater. Chem*., 1995, 5, 1919-1926.
[63] Croguennec, L; Deniard, P; Brec, R; Biensan, P; Broussely, M. *Solid State Ionics.,* 1996, 89, 127-137.
[64] Jang, YI; Huang, B; Wang, H; Sadoway, DR; Chiang, YM. *ECS Meeting Abstracts,* Vol. MA 98-2 1998, 132.
[65] Sharma, P; Moore, G; Zhang, F; Zavalij, PY; Whittingham, MS. *Electrochem. Solid-State Lett*., 1999, 2, 494-496.
[66] Shaju, KM; Rao, GVS; Chowdari, BVR. *Electrochim. Acta.,* 2003, 48, 1505-1514.
[67] Yoon, WS; Grey, CP; Balasubramanian, M; Yang, XQ; Fischer, DA; Mcbreen, G. *Electrochem. Solid-State Lett*., 2004, 7, A53-A55.
[68] Rossen, E; Jones, CDW; Dahn, JR. *Solid State Ionics.,* 1992, 57, 311-318.
[69] Spahr, ME; Novak, P; Schnyder, B; Haas, O; Nesper, R. *J. Electrochem. Soc*., 1998, 145, 1113-1121.
[70] Reed, J; Ceder, G. *Electrochem. Solid-State Lett*., 2002, 5, A145-A148.
[71] Yang, XQ; McBreen, J; Yoon, WS; Grey, CP. *Electrochem. Comm*., 2002, 4, 649-654.
[72] Ohzuku, T; Makimura, Y. *Chem. Lett*., 2001, 744-745.
[73] Makimura, Y; Ohzuku, T. *J. Power Sources.,* 2003, 119, 156-160.
[74] Komaba, S; Myung, S; Hirosaki, N; Hosoya, K; Kumagai, N. *J. Power Sources.,* 2005, 146, 645-649.

[75] Wang, Y; Jiang, J; Dahn, JR. *Electrochem. Commun.*, 2007, 9, 2534-2540.
[76] Zhou, F; Zhao, X; Lu, Z; Jiang, J; Dahn, JR. *Electrochem. Solid-State Lett.*, 2008, 11, A155-A157.
[77] Liu, Z; Yu, A; Lee, JY. *J. Power Sources.,* 1999, 81, 416-419.
[78] Kim, JM; Chung, HT. *Electrochim. Acta.,* 2004, 49, 937-944.
[79] Shaju, KM; Rao, GVS; Chowdari, BVR. *Electrochim. Acta.,* 2002, 48, 145-151.
[80] Park, SH; Yoon, CS; Kang, SG; Kim, HS; Moon, SI; Sun, YK. *Electrochim. Acta.,* 2004, 49, 557-563.
[81] Ohzuku, T; Makimura, Y. *Chem. Lett.*, 2001, 642-643.
[82] Yabuuchi, N; Ohzuku, T. *J. Power Sources.,* 2003, 119, 171-174.
[83] Myung, S; Lee, M; Komaba, S; Kumagai, N; Sun, Y. *Electrochim. Acta.,* 2005, 50, 4800-4806.
[84] Hwang, BJ; Tsai, YW; Carlier, D; Ceder, G. *Chem. Mater.*, 2003, 15, 3676-3682.
[85] Myung, ST; Komaba, S; Kumagai, N. *Solid State Ionics.,* 2004, 170, 139-144.
[86] Lu, Z; MacNeil, DD; Dahn, JR. *Electrochem. Solid-State Lett.*, 2001, 4, A191-A194.
[87] Kim, J; Kumagai, N; Chung, H. *Electrochem. Solid-State Lett.*, 2006, 9, A494-A498.
[88] Johnson, CS; Kim, JS; Lefief, C; Li, N; Vaughey, JT; Thackeray, MM. *Electrochem. Commun.*, 2004, 6, 1085-1091.
[89] Wu, Y; Manthiram, A. *Electrochem. Solid-State Lett.*, 2006, 9, A221-A224.
[90] Kim, JH; Park, CW; Sun, YK. *Solid State Ionics.,* 2003, 164, 43-49.
[91] Zheng, JM; Zhang, ZR; Wu, XB; Dong, ZX; Zhu, Z; Yang, Y. *J. Electrochem. Soc.*, 2008, 155, A775-A782.
[92] Zheng, JM; Li, J; Zhang, ZR; Guo, XJ; Yang, Y. *Solid State Ionics.,* 2008, 179, 1794-1799.
[93] Goodenough, JB. *J. Power Sources.,* 2007, 174, 996-1000.
[94] Hwang, KT; Um, WS; Lee, HS; Song, JK; Chung, KW. *J. Power Sources.,* 1998, 74, 169-174.
[95] Thackeray, MM. *Progress In Solid State Chemistry.,* 1997, 25, 1-71.
[96] Shaju, KM; Bruce, PG. *Chem. Mater.*, 2008, 20, 5557-5562.

[97] Thackeray, MM; David, WIF; Bruce, PG; Goodenough, JB. *Mater. Res. Bull.*, 1983, 18, 461-472.

[98] David, W; Thackeray, MM; De-Picciotto, LA; Goodenough, JB. *J. Solid State Chem.*, 1987, 67, 316-323.

[99] Tarascon, JM; Wang, E; Shokoohi, FK; McKinnon, WR; Colson, S. *J. Electrochem. Soc.*, 1991, 138, 2859-64.

[100] Pereamuage, D; Abraham, KM. *J. Electrochem Soc.*, 1998, 145, 1131-1136.

[101] Takada, T; Hayakawa, H; Enoki, H; Akiba, E; Slegr, H; Davidson, I; Murray, J. *J. Power Sources.,* 1999, 81, 505-509.

[102] Guyomard, D; Tarascon, JM. *J. Electrochem. Soc.*, 1992, 139:937-948.

[103] GuyomarSoc. D; Tarascon, M. *Solid State Ionics.,* 1994, 69, 222-237.

[104] Amine, K; Liu, J; Kang, S; Belharouak, I; Hyung, Y; Vissers, D; Henriksen, G. *J. Power Sources.,* 2004, 129, 14-19.

[105] Amatucci, GG; Pereira, N; Zheng, T; Tarascon, JM. *J. Electrochem. Soc.*, 2001, 148, A171-A182.

[106] Son, JT; Kim, HG; Park, YJ. *Electrochim. Acta.,* 2004, 50, 453-459.

[107] Kannan, AM; Manthiram, A. *Electrochem. Solid-State Lett.*, 2002, 5, A167-A169.

[108] Lee, SW; Kim, KS; Moon, HS; Kim, HJ; Cho, BW; Cho, WI; Ju, JB; Park, JW. *J. Power Sources.,* 2004, 126, 150-155.

[109] Sun, YK; Lee, YS; Yoshio, M; Amine, K. *Electrochem. Solid-State Lett.*, 2002, 5, A99-A102.

[110] Thackeray, MM; Johnson, CS; Kim, JS. *Electrochem. Commun.*, 2003, 5, 752-758.

[111] Park, SC; Han, YS; Kang, YS; Lee, PS; Ahn, S; Lee, HM; Lee, JY. *J. Electrochem. Soc.*, 2001, 148, A680-A686.

[112] Cho, J; Kim, GB; Lim, HS; Kim, CS; Yoo, SI. *Electrochem. Solid-State Lett.*, 1999, 2, 607-609.

[113] Liu, Z; Wang, H; Fang, L; Lee, JY; Gan, LM. *J. Power Sources.,* 2002, 104, 101-107.

[114] Padhi, AK; Nanjundaswamy, KS; Goodenough, JB. *Solid State Ionics.,* 1996, 92, 1-10.

[115] Padhi, AK; Nanjundaswamy, KS; Masquelier, C; Goodenough, JB. *J. Electrochem Soc.*, 1997, 144, 2581-2586.

[116] Padhi, AK; Nanjundaswamy, KS; Goodenough, JB. *J. Electrochem Soc.*, 1997, 144, 1188-1194.

[117] Padhi, AK; Nanjundaswamy, KS; Masquelier, C; Okada, S; Goodenough, JB. *J. Electrochem Soc.*, 1997, 144, 1609-1613.
[118] Tarascon, JM; Armand, M. *Nature.,* 2001, 414, 359-367.
[119] Prosini, PP; Zane, D; Pasquali, M. *Electrochim. Acta.,* 2001, 46, 3517-3523.
[120] Yamada. A; Chung, SC; Hinokuma, K. *J. Electrochem. Soc.*, 2001, 148, A224-A229.
[121] Li, G; Azuma, H; Tohda, M. *J. Electrochem. Soc.*, 2002, 149, A743-A747.
[122] Chung, SY; Bloking, JT; Chiang, YM. *Nature Mater.*, 2002, 1, 123-128.
[123] Yang, S; Song, Y; Zavalij, PY; Whittingham, MS. *Electrochem. Comm.*, 2002, 4, 239-244.
[124] Chen, Z; Dahn, JR. *J. Electrochem. Soc.*, 2002, 149, A1184-A1189.
[125] Yang, S; Song, Y; Ngala, K; Zavalij, PY; Whittingham, MS. *J.Power Sources.,* 2003, 119, 239-246.
[126] Andersson, AS; Kalska, B; Haggstrom, L; Thomas, JO. *Solid State Ionics.,* 2000, 130, 41-52.
[127] Arnold, G; Garche, J; Hemmer, R; Ströbele, S; Vogler, C; Wohlfahrt-Mehrens, M. *J. Power Sources.,* 2003, 119, 247-251.
[128] Andersson, AS; Thomas, JO; Kalska, B; Haggstrom, L. *Electrochem Solid-State Lett.*, 2000, 3, 66-68.
[129] Takahashi, M; Tobishima, S; Takei, K; Sakurai, Y. *Solid State Ionics.,* 2002, 148, 283-289.
[130] Tukamoto, H; West, AR. *J. Electrochem. Soc.*, 1997, 144, 3164-3168.
[131] Guan, J; Liu, M. *Solid State Ionics.,* 1998, 110, 21-28.
[132] Prosini, PP; Lisi, M; Zane, D; Pasquali, M. *Solid State Ionics.,* 2002, 148, 45-51.
[133] Rissouli, K; Benkhouja, K; Ramos-Barrado, JR; Julien, C. *Mater. Sci. Eng. B.*, 2003, 98, 185-189.
[134] Franger, S; Le Cras, F; Bourbon, C; Rouault, H. *J. Power Sources.*, 2003, 119, 252-257.
[135] Higuchi, M; Katayama, K; Azuma, Y; Yukawa, M. Suhara, M. *J. Power Sources.,* 2003, 119, 258-261.
[136] Barker, J; Saidi, MY; Swoyer, JL. *Electrochem. Solid-State Lett.*, 2003, 6, A53-A55.
[137] Yang, S; Zavalij, PY; Whittingham, MS. *Electrochem. Commun.*, 2001, 3, 505-508.

[138] Hu, Y; Doeff, MM; Kostecki, R; Finones, R. *J. Electrochem. Soc.*, 2004, 151, A1279-1285.

[139] Croce, F; D'Epifanio, AD; Hassoun, J; Deptula, A; Olczac, T; Scrosati, BJ. *Electrochem. Solid-State Lett.*, 2002, 5, A47-A50.

[140] Huang, H; Yin, S; Nazar, LF. *Electrochem. Solid-State Lett.*, 2001, 4, A170-A172.

[141] Chen, Z; Dahn, JR. *J. Electrochem. Soc.*, 2002, 149, A1184-A1189.

[142] Okada, S; Sawa, S; Egashira, M; Yamaki, J; Tabuchi, M; Kageyama, H; Konishi, T; Yoshino, A. *J. Power Sources.*, 2001, 97-8, 430-432.

[143] Amine, K; Yasuda, H; Yamachi, M. *Electrochem. Solid-State Lett.*, 2000, 3, 178-179.

[144] Li, G; Azuma, H; Tohda, M. *Electrochem. Solid-State Lett.*, 2002, 5, A135-A137.

[145] Zhou, F; Cococcioni, M; Kang, K; Ceder, G. *Electrochem. Comm.*, 2004, 6, 1144-1148.

[146] Liu, J; Manthiram, A. *J. Electrochem. Soc.*, 2009, 156, A66-A72.

[147] Morgan, D; Van der Ven, A; Ceder, G. *Electrochem. Solid-State Lett.*, 2004, 7, A30-A32.

[148] Delacourt, C; Poizot, P; Morcrette, M; Tarascon, JM; Masquelier, C. *Chem. Mater.*, 2004, 16, 93-99.

[149] Fang, H; Pan, Z; Li, L; Yang, Y; Yan, G; Li, G. Wei, S. *Electrochem. Comm.*, 2008, 10, 1071-1073.

[150] Drezen, T; Kwon, N; Bowen, P; Teerlinck, I; Isono, M; Exnar, I. *J. Power Sources.*, 2007, 174, 949-953.

In: Lithium Batteries: Research, Technology… ISBN: 978-1-60741-722-4
Editors: G. R. Dahlin and K. E. Strom

Chapter 3

ANALYSIS OF CELL IMPEDANCE FOR THE DESIGNOF A HIGH-POWER LITHIUM-ION BATTERY

Hyung-Man Cho and Heon-Cheol Shin*
School of Materials Science and Engineering,
Pusan National University, Busan, Korea.

ABSTRACT

This work presents a systematic semi-empirical way to analyze the constituents of total cell impedance in a lithium-ion battery, and their time-dependent contributions to total direct current (dc) polarization. The approach includes the differentiation of internal resistive elements, followed by theoretical calculations of their contributions to total polarization using circuit analysis. Our method provides a fast and reliable way to design a high-power battery with the instantaneous input/output power that best fits the user's specific needs. It also provides insight into the design of high-power with long shelf life and calendar life. We begin with an overview of high-power cell design. Methodology to differentiate and quantify the time-dependent contribution of elementary resistances to total polarization is given, and applications to power aging in battery use, and power decline at low operating temperature, are

---

* Corresponding author: E-mail: hcshin@pusan.ac.kr

suggested. A strategy for the design of materials to meet power requirements is discussed for each case.

I. INTRODUCTION

High power batteries have been widely studied as an energy source for eco-friendly transportation systems including hybrid electric vehicles, electric motors, ships, and aircraft that can be operated with little or no oil consumption and carbon dioxide emissions, as compared to conventional systems with an internal combustion engine [1-3]. The lithium-ion battery (LIB) has been one of the most promising substitutes for the nickel metal hydride battery used in most of the hybrid electric vehicles (HEVs) commercially available today [4].

A successful design of a high-power LIB at both ambient and low operating temperatures, the diagnosis of the battery with degraded power, and the subsequent redesign of the battery with long shelf/calendar life rests entirely on an in-depth understanding of factors related to battery power. This chapter presents a systematic semi-empirical way to analyze the time-dependent contribution to total polarization of each reaction step involved in battery operation. For this purpose, electrical signals from the cathode and anode are separated using a three-electrode electrochemical cell configuration. Then, they are further differentiated on the grounds of mechanism-based or phenomenological equivalent circuits. Next, the variation in elementary polarization caused by each reaction step with time is calculated using a theoretical analysis of an equivalent circuit. This gives the proportional contribution of each reaction step to the total polarization during a pulse discharging or regenerative (charging) process.

In this article, the main impedance factors affecting the power performance of a LIB are proposed as functions of operating temperatures and degrees of cell power degradation. In Section II, high power cell design is introduced with an emphasis on design factors of the constituents. In Section III, we suggested the methodology to quantitatively analyze the time-dependent contribution of elementary resistances to total polarization. In Section IV, an in-depth diagnosis of the battery with degraded power is considered. In particular, analyses based on two- and three-electrode electrochemical cell configurations are compared. Finally, Section V describes the temperature dependence of battery power and related critical factors, together with a prediction of the low temperature performance of a hybrid electrode consisting of typical insertion materials and electrical double layer capacitor (EDLC) materials.

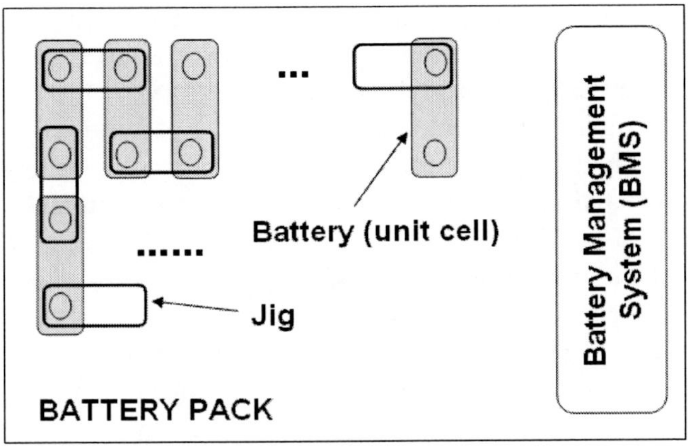

(a)

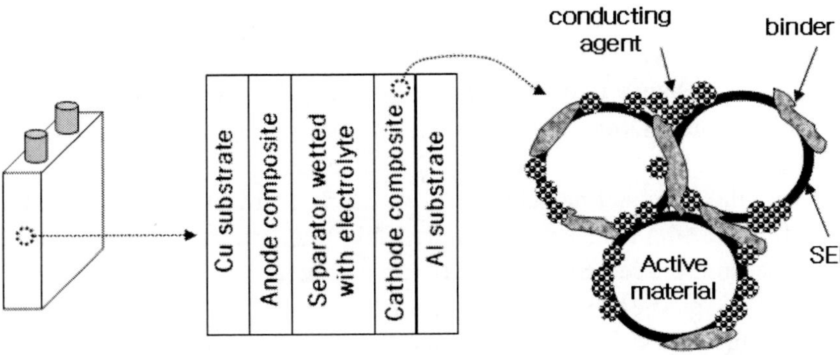

(b)

Figure 1. Schematic representations of (a) battery pack and (b) the inside of unit cell.

II. Overview of High Power Cell Design

The main role of a LIB in transportation systems such as a HEV depends on the type of electric vehicle. Power density is particularly important when the battery is used as an auxiliary power source for a strong (full) or plug-in HEV [5]. Here, the engine might be operated at the most efficient speed and additional power is supplied by the battery as required. That is, the battery provides additional power in the case of uphill driving and accelerating, and absorbs regenerative power during downhill/normal driving and braking. Accordingly, the battery, as an assisting power source, needs superior capability for charge (or power) release and gain.

The successful development of a LIB with high power density depends on minimizing the ionic and electronic resistances related to battery operation. Thus, it is imperative to understand the pathways of electrons and lithium ions. The resistance of the battery or unit cell (hereafter called "cell resistance") is only a part of the total resistance responsible for battery power. The design must take into consideration battery pack resistance; this consists of cell resistance, the bulk resistance of jigs, the connection (or contact) resistance between cell and jig, the resistance caused by the battery management systems (BMS), *etc.*, as shown in Figure 1(a). Among these resistance sources, cell resistance is of prime importance because the consistency in cell resistance during and after repetitive use (or long time storage), and its temperature dependence especially at very low temperatures, are decisive factors in the performance of the whole battery pack.

A simplified schematic view inside a LIB is shown in Figure 1(b). The internal cell resistance in the LIB can be roughly classified as follows [6-8]: 1) a series resistance R_Ω due to lithium ion transport through the porous separator wetted with the electrolyte, and electron transport through numerous pathways including the conducting chains (*e.g.*, carbon black in the composite electrode); 2) a film resistance R_f due to lithium ion transport through the solid electrolyte interphase (SEI) on the active materials; 3) a charge transfer resistance R_{ct} at the interface between the electrolyte (or SEI) and active materials; and 4) a diffusion impedance Z_{diff} against solid-state bulk diffusion of lithium ions through active materials.

The design to decrease R_Ω is straightforward in that the electronic or ion conductivity of the components has a direct influence on the value of R_Ω. For instance, separators with different porosity and pore tortuosity lead to different values of apparent electrolyte resistance, since these two factors play a critical role in the polarization inside the separator. (In fact, air permeability and the Gurley number are more frequently used in industry to evaluate the degree of penetrability of active species through the separator than the terminologies such as porosity, tortuosity, and polarization [9].) The separators with more porous and less tortuous pore structure result in smaller R_Ω (or polarization), yielding higher power density of the battery. However, an excessively open separator pore structure is usually not adequate for practical use due primarily to safety issues and a strong possibility of micro-shorts between the cathode and anode [10].

Although the chemical composition of the SEI has been extensively studied for two decades, a systematic analysis of the lithium ion transport through it has not been performed in spite of its relevance. This is because the unique structure of the SEI, consisting of a compact Li_xPF_y/Li_2CO_3-based inner layer and porous polycarbonate-based outer layer, complicates the situation. The control of the SEI formation (or activation) process immediately after battery assembly, and the appropriate design of an electrolyte with specific additives that are decomposed ahead of the conventional SEI-forming solvent such as ethylene carbonate may affect SEI properties, but a study of the design parameters of the SEI is still a great challenge [11].

Energetically, interfacial charge transfer and solid-state lithium diffusion through the active materials are usually the most difficult reaction steps in the lithium intercalation/deintercalation process. In particular, R_{ct} is thought to be critical in battery performance at very low temperatures although there is still controversy regarding the dominant low temperature rate-determining mechanism [12-14]. Design of the composition of active materials and surface modification would be promising ways to elevate electrochemical activity at the interface, thereby dropping off the charge transfer barrier. Nano-structured active materials with nano-pores and/or particles are an alternative with great potential for significantly lowering the value of R_{ct}. However, the difficulty of uniform dispersion, and the irreversible charge consumption during the initial SEI formation process due to their extremely large surface area, needs to be resolved for their practical use [15-17].

The solid-state diffusion process is particularly important when interfacial charge transfer is relatively facile and thus the total intercalation reaction is governed by the chemical diffusion of lithium through the active materials. Basically, the activation energy for lithium diffusion closely relates to the size of the lithium diffusion path and the electrostatic interaction between lithium ions and the cations around it. Hence, much work has been done to reduce the activation energy for lithium diffusion by compositional and/or structural adjustment at the atomic scale on the basis of *ab initio* calculation [18, 19]. At the microscale, the modification of microstructure such as grain size (or grain boundary) is expected to considerably affect lithium diffusivity. In general, the oxides with larger primary particle sizes show higher lithium diffusivity since the inter-particle transport of lithium across grain boundaries is reduced [20].

III. Time-Dependent Contribution of Reaction Steps to Total Polarization

1. Overview of the Approach

Our goal is to quantify the time-dependent proportional contribution of individual reaction steps to total polarization during battery operation, and then set the redesign strategy of the cell components of a high-power LIB. The differentiation and quantification of the elementary impedances are the first step in this approach. Here, the electrochemical impedance spectra are obtained at a specific state-of-charge (SOC) and battery operating temperature. Then, the lithium intercalation is modeled to construct a mechanism-based equivalent circuit, and its electrical parameters are estimated using a complex nonlinear least squares (CNLS) fitting method [21, 22]. If the reliable identification of an impedance model and its electrical parameters is unlikely due to ambiguity of the measured impedance spectra, impedance diagnosis methods such as discrete Fourier transformation (DFT) [23-26] and differential impedance analysis (DIA) [27-29] could be used to analyze the impedance spectra without prior assumption regarding individual reactions.

The next step is the calculation of elementary polarizations due to the corresponding resistive elements as a function of pulse discharging/charging time, with the help of theoretical analysis of an equivalent circuit. The Simulation Program with Integrated Circuit Emphasis (SPICE) software program is an excellent circuit simulation tool [30-32]. With the help of SPICE, the variation of instantaneous elementary polarization with battery operating time is attainable by considering the potential difference between the nodes of both sides of the resistive elements of concern.

Then, time-dependent proportional contributions of elementary polarizations (or reaction steps) to total dc polarization are finally determined, and this gives useful information when we design the battery with the instantaneous power density that fits our specific needs. For example, if the HEV to be produced typically undergoes repetitive short time acceleration and deceleration (braking), cell constituents need to be designed for enhanced power performance for the initial time period of battery discharging and charging at high current (or power) density. On the other hand, when the driving pattern includes extended acceleration or uphill driving, power performance over a prolonged time period must be considered and the cell constituents designed accordingly. In addition to cell design according to driving patterns, this approach can be effectively utilized for diagnosis of a battery with degraded power, analysis of low temperature power decline, and prediction of power performance. The application of this approach to a hypothetical electrochemical cell is given in the next section.

2. Model Case: Analysis on Hypothetical Electrode in LIB

It is generally accepted that the lithium intercalation/deintercalation process of the LIB consists of Li^+ transport through the electrolyte, Li^+ transport through the SEI coupled with the charge storing in it, interfacial charge transfer combined with electrical double layer charging, and solid-state lithium diffusion into the active material [6-8]. Figure 2 shows a typical equivalent circuit that models the lithium intercalation/deintercalation process.

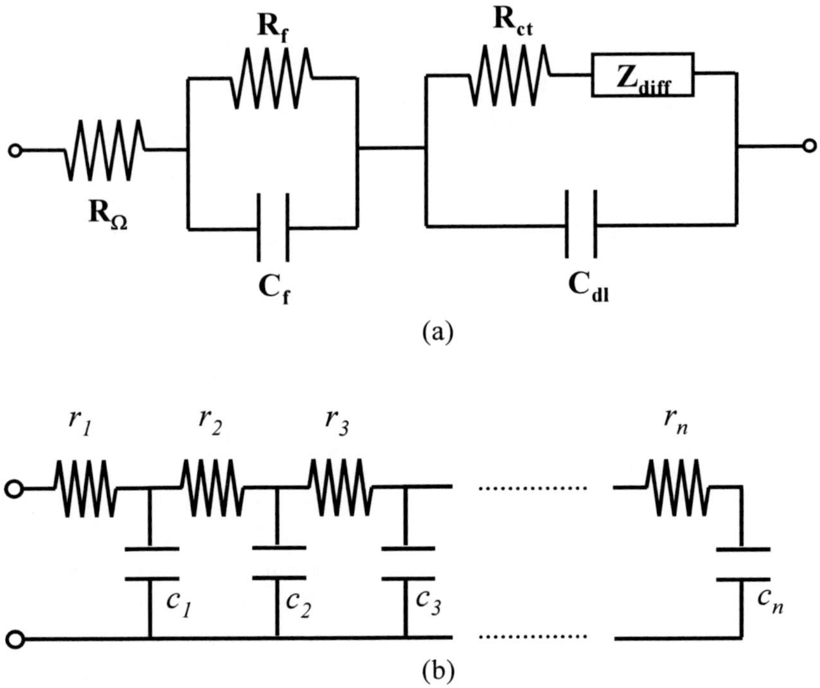

Figure 2. (a) Simplified hypothetic equivalent circuit to model the intercalation process of LIB and (b) the electrical expression of the Warburg or diffusion impedance, Z_{diff}, *i.e.*, transmission line (TML).

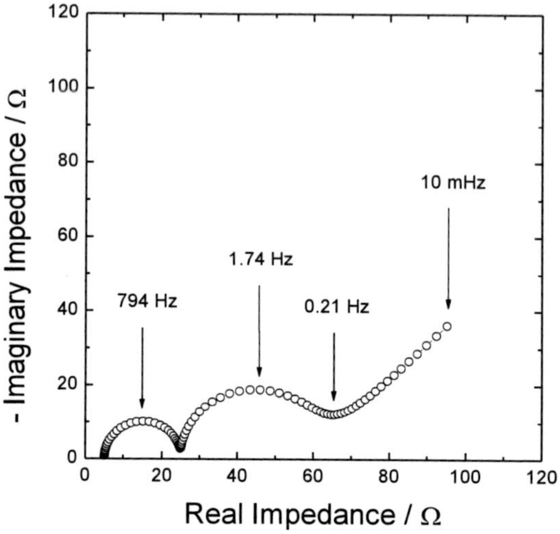

(a)

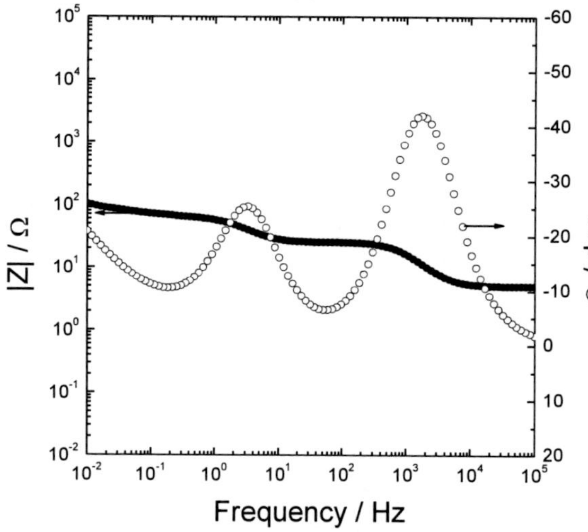

(b)

Figure 3. (a) Nyquist plot and (b) Bode plot, obtained from the equivalent circuit of Figure 2. The impedance spectra were theoretically determined by arbitrarily taking $R_{\Omega}=5\ \Omega$, $R_f=20\ \Omega$, $C_f=10\ \mu F$, $R_{ct}=35\ \Omega$, and $C_{dl}=2\ mF$. The diffusion impedance Z_{diff} is expressed as $Z_{diff}=A_{\omega}(j\omega)^{-0.5}tanh[\delta(j\omega)^{0.5}]$ (where, δ is defined as $L/D^{1/2}$, ω is the angular frequency, and A_{ω} is the Warburg coefficient expressed as R_D/δ). $R_D=400\ \Omega$, $L=10\ \mu m$, and $D=10^{-9}\ cm^2/s$ were taken for the calculation of Z_{diff}. The elemental resistance r_n and capacitance c_n in the TML were estimated to be $4\times10^7\ \Omega\cdot m^{-1}$ and $2.5\times10^5\ s\cdot\Omega^{-1}\cdot m^{-1}$, respectively.

The electrical parameters related to electrode reaction were arbitrarily assigned to construct the hypothetical electrode in a LIB. In particular, for the solid-state diffusion process, finite-length lithium diffusion through the active materials was assumed. Figure 3 shows the impedance spectra of a hypothetical electrode in the frequency range of 1 MHz to 1 mHz (for details on the electrical parameters, see the caption of Figure 3). In particular, the Nyquist plot shown in Figure 3(a) has two semicircles in the high and intermediate frequency range, followed by a 45° inclined line at the low frequency range. The first and second semicircles are ascribed to the reactions in the SEI and the interfacial charge transfer reaction combined with electrical double layer charging, respectively, while the inclined line is due to solid-state lithium diffusion into the active material.

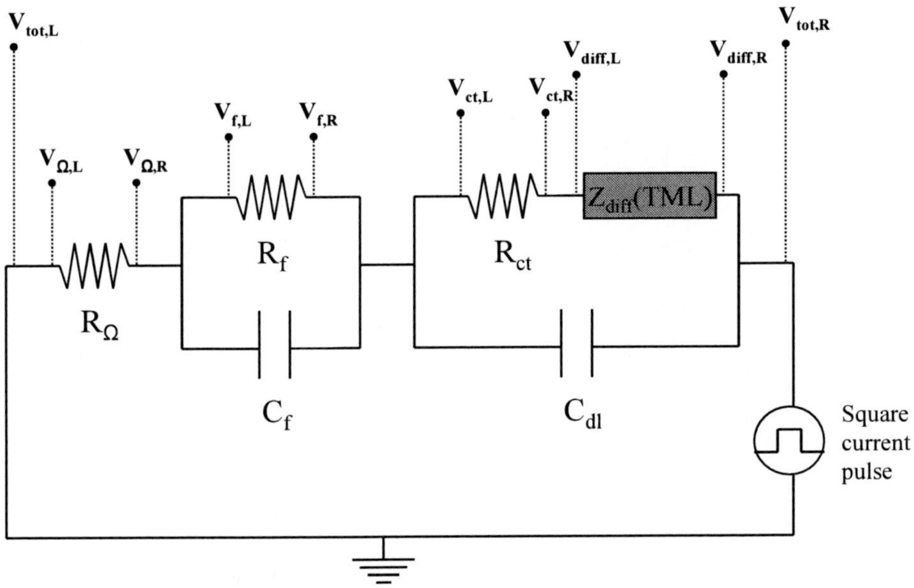

Figure 4. Brief description of the circuit analysis, showing the application of the square current pulse and the estimation of the potential difference, *i.e.*, polarization, due to the electrical elements. Potential differences between the nodes, $V_{\Omega L}$ / $V_{\Omega R}$, $V_{f,L}$ / $V_{f,R}$, $V_{ct,L}$ / $V_{ct,R}$, $V_{diff,L}$ / $V_{diff,R}$, *and* $V_{tot,L}$ / $V_{tot,R}$ represent the polarizations due to the uncompensated Ohmic resistance (ΔV_{Ω}), resistance to lithium-ion transport through the SEI layer (ΔV_f), interfacial charge-transfer resistance (ΔV_{ct}), diffusion resistance (ΔV_{diff}), and total cell resistance (ΔV_{tot}), respectively.

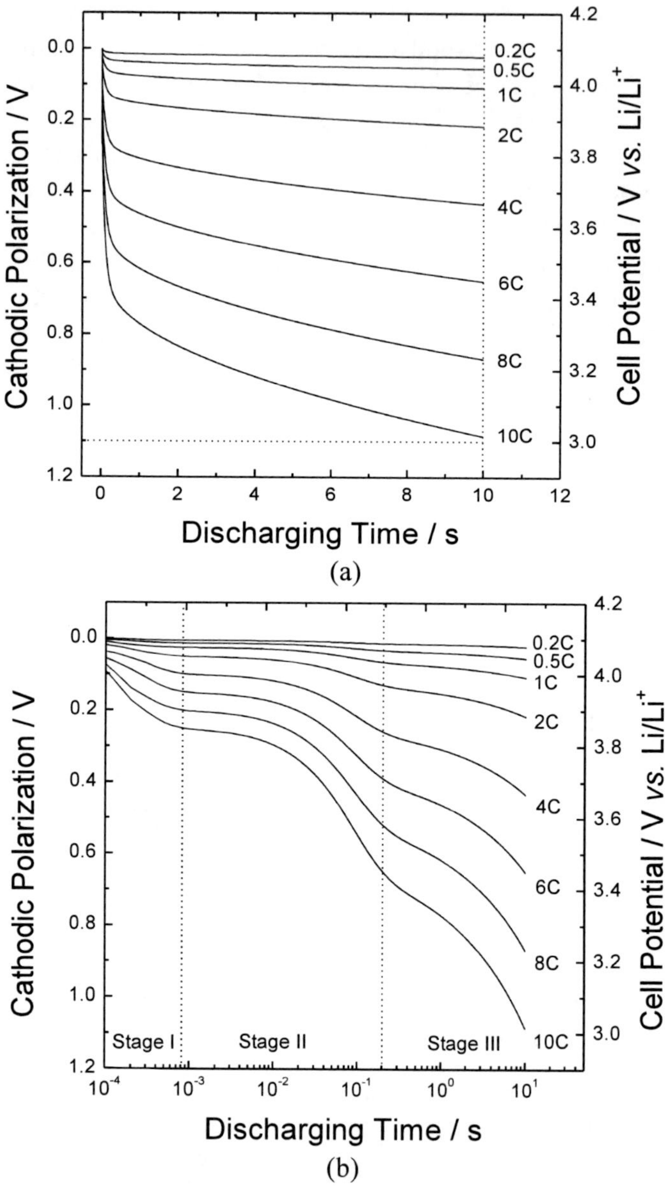

Figure 5. (a) Calculated cathodic polarization (or cell potential) transients during cathodic pulse discharging for 10 s at different rates and (b) the reproduced plot in a semi-logarithmic scale which shows three-stage behaviour. In order to convert the current density to the C rate, the mass of the active material and its specific gravimetric capacity were assumed to be 10 mg and 100 $mAhg^{-1}$, respectively.

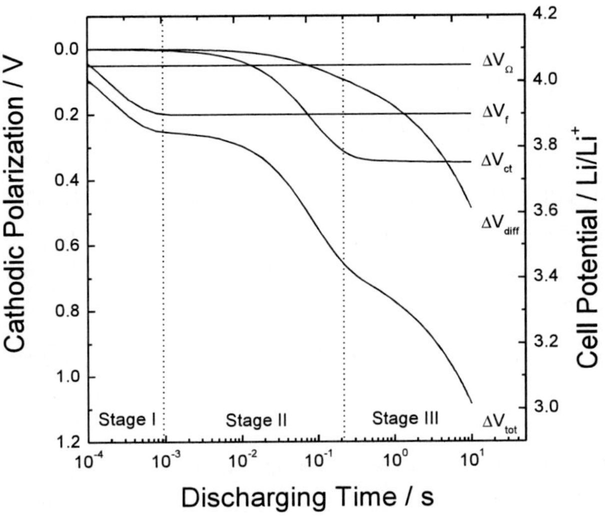

Figure 6. Variation of the elementary polarizations with the discharging time at a discharging rate of 10C on a semi-logarithmic scale, determined from the circuit analysis (Figure 4).

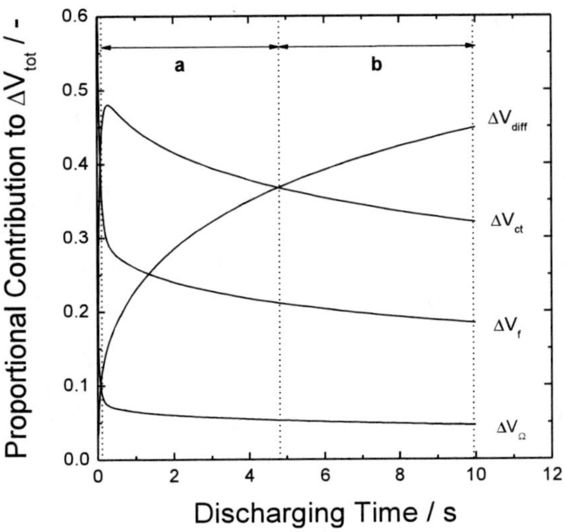

Figure 7. Time-dependent proportional contribution of elementary impedances to total polarization, reproduced from Figure 6. The polarizations due to the charge transfer and the diffusion make a maximum contribution to total polarization in the regions of ***a*** and ***b***, respectively.

For a theoretical analysis of an equivalent circuit based on SPICE, the Warburg or diffusion impedance Z_{diff} (Figure 2(a)) was expressed using a combination of the electrical elements. A resistive-capacitive transmission line might be an excellent electrical analogy of the ideal diffusion process (Figure 2(b)) [33]. The elemental resistance r_n and capacitance c_n of the transmission line have a direct relationship with the limiting diffusion resistance R_D ($R_D = r_n \cdot L$, where L is the diffusion thickness) and diffusion capacitance C_D ($C_D = c_n \cdot L = L^2/[R_D \cdot D]$, where D is the chemical diffusion coefficient), respectively, in the finite length diffusion process. Since R_D is correlated with the Warburg coefficient A_ω ($A_\omega = R_D \sqrt{D} / L$), which can be experimentally estimated from the impedance spectrum in the low frequency region, the evaluation of r_n and c_n is quite straightforward [33-35].

Now, different current pulses are applied to the reconstructed equivalent circuit where the Warburg element is substituted for the transmission line, as shown in Figure 4. The calculation of the potential difference between the nodes of $V_{tot,L}$ and $V_{tot,R}$ as a function of current application time generates the time-dependent total polarization. We note that the resulting total polarization is monotonically decreasing on a linear scale (Figure 5(a)), while it shows three-stage behavior on a semi-logarithmic scale (Figure 5(b)), strongly indicating the change in the governing factor in the shape of total polarization transients.

The complicated shape of semi-logarithmic polarization transients can be readily understood by considering the variations of elementary polarizations with time. The time-dependent elementary polarizations were obtained for the nodes of $V_{\Omega L}/V_{\Omega R}$ (for series resistance, including electrolyte resistance), $V_{f,L}/V_{f,R}$ (for SEI film resistance), $V_{ct,L}/V_{ct,R}$ (for interfacial charge transfer resistance), and $V_{diff,L}/V_{diff,R}$ (for bulk diffusion resistance). The resulting elementary polarization transients at a current of 10 mA (Figure 6) proved that the polarizations due to SEI film resistance (ΔV_f), interfacial charge transfer resistance (ΔV_{ct}), and solid-state diffusion resistance (ΔV_{diff}) are attributable, consecutively, to three-stage behavior of the total polarization transients.

The contribution of the individual reaction steps to total polarization was calculated from Figure 6, which gives us the reaction step (or elementary polarization) corresponding to the maximum contribution to total polarization at a specific moment of discharging time. In our hypothetical electrode, the interfacial charge transfer reaction primarily affects total polarization over an initial period of discharging time, while the solid-state diffusion process makes a maximum contribution in a later stage of discharging time, as shown in

Figure 7. That is, to enhance the initial power performance of our hypothetical electrode, the surface design of active materials to elevate the electrochemical activity must come before all other design of the cell constituents. On the other hand, when the power performance must not drop off significantly in the later stage of discharging time and level off throughout the whole discharging time, the bulk design of active materials to promote solid-state diffusion takes precedence over all other design factors.

Here, following three factors to additionally affect the cell potential during discharging process should be mentioned: As a matter of fact, all the resistances and capacitances except for a series resistance R_{Ω} in the circuit are basically dependent on lithium content in the active materials. In particular, it has been well known that R_{ct} and Z_{diff} are strong functions of lithium content. This indicates that their variations with lithium content needs to be taken into account for the calculation of the polarization. Furthermore, the change in the electrochemical (thermodynamic) potential with lithium content, which relates to the absorption isotherm of the systems, should be considered for the calculation. Nevertheless, in the relatively mild operating conditions adopted in this work, *i.e.*, the maximum discharging current of less than 15C rate and discharging time for less than 10 s, the variation of lithium content (stoichiometry) is estimated to be less than 0.02, which is quite small so that the change in resistance/capacitance/diffusivity and cell potential with lithium content could be virtually disregarded for the calculation of the potential transients [31]. In the case of extremely high current drains, however, the above two factors can't be neglected. Finally, the difficulty in lithium ion transport through the porous separator wetted with the electrolyte might raise the polarization. This additional polarization is usually unavoidable at the extremely high current drains and get larger as one uses separators with smaller penetrability, *i.e.*, less porous and more tortuous separators [9].

IV. In-Depth Diagnosis of the Battery with Degraded Power

1. Cell Configuration and Electrochemical Test Procedures

Two-electrode and three-electrode electrochemical cells used for the electrochemical measurements are shown in Figures 8(a) and (b), respectively. In both cell configurations, lithium foil is selected as the counter (auxiliary) electrode. Composite materials of 90 wt.% $LiCoO_2$ (Aldrich), 5 wt.% carbon black, and 5 wt.% polyvinylidene fluoride binder in n-methyl pyrrolidinone were employed as the working electrode. Details of the drying and pressing conditions of the composite electrode can be found in [32]. A Celgard 2400 separator was wetted with 1 M solution of lithium hexafluorophosphate ($LiPF_6$) in a 1:1 volume mixture of ethylene carbonate (EC) and diethyl carbonate (DEC), and then sandwiched between a composite working electrode and a counter electrode.

For the reference electrode of a three-electrode electrochemical cell, the end tip of the Teflon coated copper wire was removed. Lithium titanium oxide ($Li_4Ti_5O_{12}$, Altair) was coated on the bare copper [36, 37]. The reference electrode was located between two separators, as shown in Figure 8(b). Before conducting the electrochemical tests, the state of charge of the lithium titanium oxide was set to 50%, and in this state its potential was approximately 1.57 V (*vs.* Li/Li^+). All of the cells were assembled in a glove box (MBraun, Germany) filled with purified argon gas.

Tests of two-electrode electrochemical cell: The as-prepared cell was first galvanostatically activated five times between 3.0 and 4.2 V (*vs.* Li/Li^+) at a rate of 0.2 C (24 mAg^{-1}; a gravimetric specific capacity of 120 $mAhg^{-1}$ was assumed to convert the current density into the C rate). The cell in this state is hereafter called the "fresh cell." After the impedance measurement and a series of current pulse tests, the cell was further galvanostatically cycled 20 times under the same conditions that the as-prepared cell was initially cycled (the cell in this state is hereafter called the "aged cell"). The specific capacity of the aged cell was reduced to about 80%, as compared to the capacity of the fresh cell. Finally, impedance measurement and current pulse tests were performed for the aged cell. For the pulse tests, after equilibrating the cell at 4.2 V (*vs.* Li/Li^+), a variety of cathodic currents were applied until either the electrode potential reached the low cutoff voltage of 3.0 V (*vs.* Li/Li^+) or the cell was discharged for 10 s. The impedance measurements were carried out at a potential of 4.2 V (*vs.* Li/Li^+) by applying an ac amplitude of 5 mV_{rms} over the frequency range of 10 mHz to 100 kHz.

Tests of three-electrode electrochemical cell: The initial activation of the as-prepared cell was done in the same way as that of the two-electrode cell. Then, the fresh cell was further cycled 20 times for cell aging. The electrochemical impedance spectra at a cell potential of 4.1 V (*vs.* Li/Li^+) were obtained for the fresh and aged cells for the frequency range of 50 kHz to 5 mHz. Current pulse tests were carried out immediately after the impedance tests by applying different cathodic currents for 10 s.

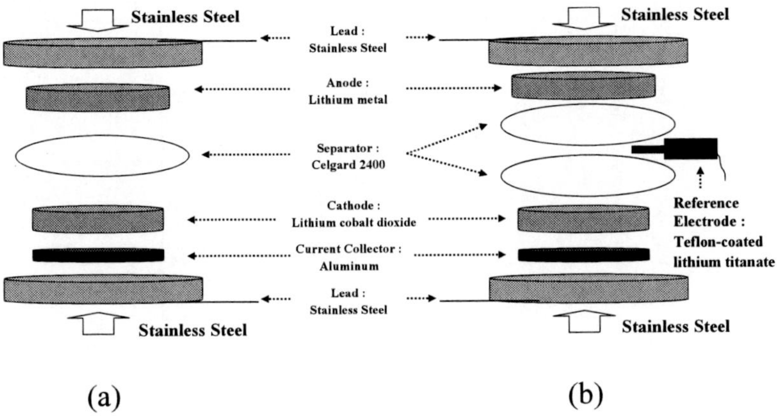

Figure 8. Schematic illustrations of (a) two-electrode and (b) three-electrode electrochemical cell configurations.

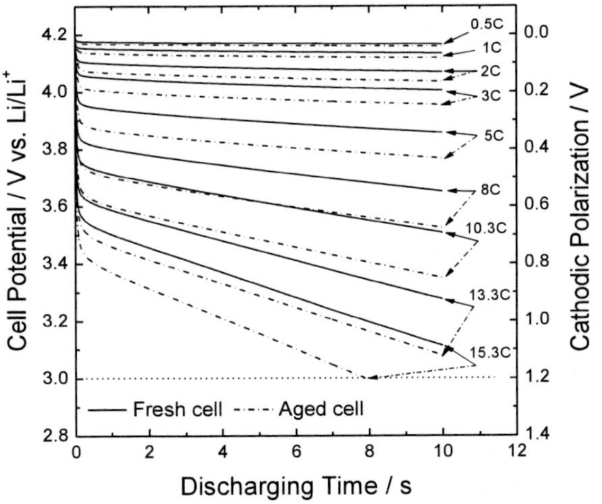

Figure 9. Typical potential (or cathodic polarization) transients during the cathodic pulse discharging for 10 s at different rates, obtained from the fresh cell (solid line) and the aged cell (dashed line). Figure 1a in D.-K. Kang and H.-C. Shin, "Investigation on cell impedance for high-power lithium-ion batteries", *Journal of Solid State Electrochemistry* 11 (2007) 1405-1410, Copyright © (2007), with kind permission of Springer Science and Business.

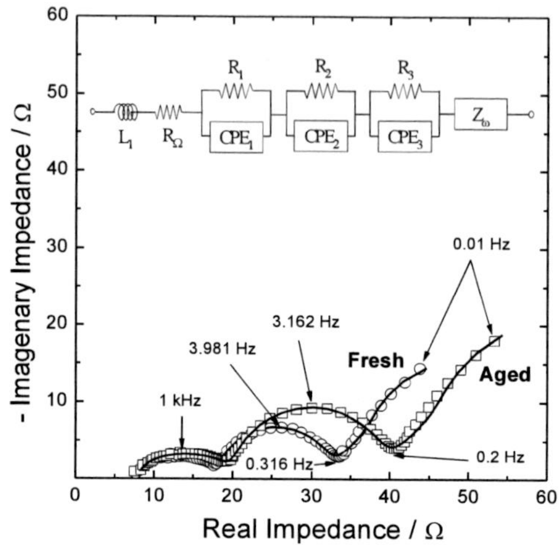

Figure 10. Typical impedance spectra, obtained from the fresh (open circle) and the aged cell (open square) at the cell potential of 4.2 V (*vs.* Li/Li^+). The *inset* is the phenomenological equivalent circuit to model the overall process. Solid lines were determined from the CNLS fittings of the impedance spectra to the equivalent circuits. Figure 2 in D.-K. Kang and H.-C. Shin, "Investigation on cell impedance for high-power lithium-ion batteries", *Journal of Solid State Electrochemistry* 11 (2007) 1405-1410, Copyright © (2007), with kind permission of Springer Science and Business.

A Solartron 1287 electrochemical interface was employed for all of the galvanostatic experiments. For the electrochemical impedance measurements, the Solartron 1287 electrochemical interface was coupled with a Solartron 1455A frequency response analyzer.

2. Analysis Based on a Two-Electrode Electrochemical Cell and its Limitation

Typical potential (or cathodic polarization) transients during constant current discharging for 10 s at different rates are shown in Figure 9. All the transients exhibited an abrupt drop in the initial stage of pulse discharging, followed by a relatively slow decrease in potential with time. The rate of increase in polarization increased with the discharging rate. In particular, we note that the maximum rate where continuous discharging for 10 s was achievable decreased from approximately 15 C in the fresh cell to 13 C in the aged cell. This strongly indicates that the cell resistance of the aged cell became larger than that of the fresh cell; *i.e.*, power performance degraded considerably after cell aging.

To investigate the elementary reaction steps and their contribution to the total intercalation process, impedance spectra for the fresh and aged cells were obtained as shown in Figure 10. The impedance spectra consisted of semicircles and an inclined line. Although a semicircle typically results from the interfacial reactions and an inclined line is attributable to the solid-state diffusion process, the overlapping of the time constants of the reactions in the anode and cathode makes reliable mechanism-based analysis of the impedance spectra quite unlikely.

Thus, the equivalent circuit for the overall process was phenomenologically constructed, as shown in the inset of Figure 10. The first two and the third parallel resistor-constant phase elements (CPEs) stand for the extremely depressed semicircles in the high frequency range and the third semicircle in the intermediate frequency range, respectively. The Warburg element Z_{ω} represents an inclined line in the low frequency range. The values of resistance, capacitance, and the Warburg coefficient were estimated using CNLS fitting methods. For the theoretical analysis of an equivalent circuit, the Warburg impedance was modeled by a transmission line, and its elementary resistance r and capacitance c were evaluated based on the method suggested in Section III-2. Here, all the CPEs were considered as purely capacitive components in spite of some discrepancies between the experimental potential transients and the calculated values. It seems that this simplification is tolerable in our case, since the capacitive element or CPE has affects the shape of the transients by less than a couple of hundred milliseconds [31, 32].

It is additionally noted that the changes in resistance/ capacitance/diffusivity and electrochemical or thermodynamic potential with lithium content were disregarded for the calculation of the polarization. This seems to be acceptable in the operating conditions adopted in this work due to small variation of lithium content during the discharging for 10 s. For further discussion on this topic, see [31].

The dependences of potential (or polarization) on discharging time were calculated using the SPICE program by applying the different values of square current to the equivalent circuit. The resulting transients of total cell polarization are depicted in Figure 11(a). The calculated transients quantitatively matched the experimental transients (Figure 9). We note that the semi-logarithmic variation of total cell polarization with discharging time clearly showed three-stage behavior, as demonstrated at the rate of 8 C in Figure 11(b). The calculated transients of the elementary polarizations (Figure

11(b)) tell us that the elementary polarizations due to the R_2-C_2, R_3-C_3, and diffusion impedance are responsible for the three-stage behavior of the total polarization transient.

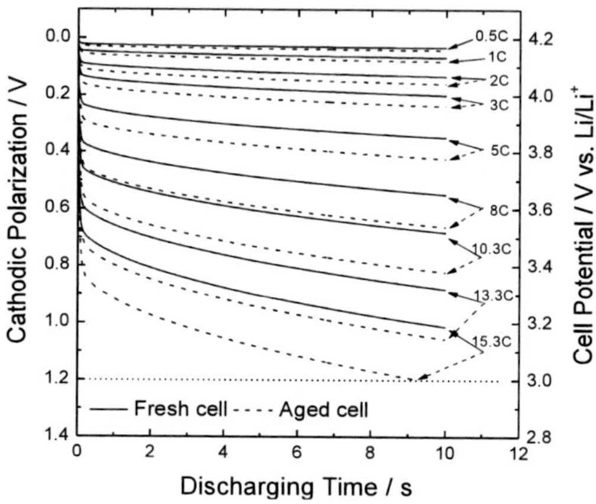

(a)

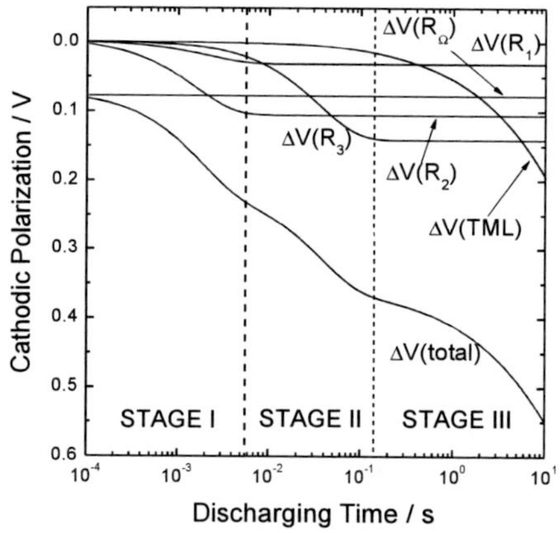

(b)

Figure 11. (a) Calculated potential (or cathodic polarization) transients during the cathodic pulse discharging for 10 s at different rates, obtained from the fresh cell (solid line) and the aged cell (dashed line) and (b) semi-logarithmic variations of the elementary and total polarizations of the fresh cell with discharging time at 8C rate, determined from the circuit analysis. Figures 3a and 4 in D.-K. Kang and H.-C. Shin, "Investigation on cell impedance for high-power lithium-ion batteries", *Journal of Solid State Electrochemistry* 11 (2007) 1405-1410, Copyright © (2007), with kind permission of Springer Science and Business.

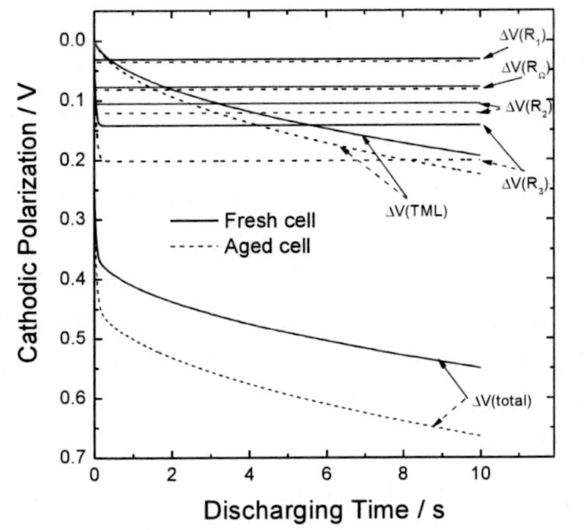

(a)

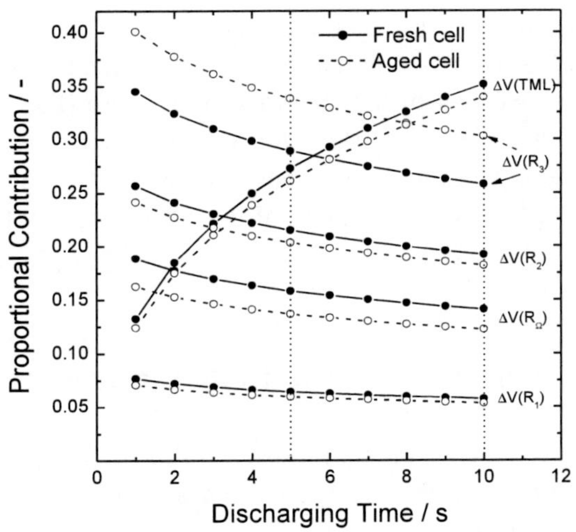

(b)

Figure 12. (a) Variations of elementary and total polarizations, obtained from the fresh cell (solid line) and the aged cell (dashed line), with discharging time at 8C rate and (b) dependence of proportional contributions of the elements to total polarization, reproduced from (a). Figures 5a and b in D.-K. Kang and H.-C. Shin, "Investigation on cell impedance for high-power lithium-ion batteries", *Journal of Solid State Electrochemistry* 11 (2007) 1405-1410, Copyright © (2007), with kind permission of Springer Science and Business.

Shown in Figure 12(a) are the calculated transients of elementary and total polarization at the rate of 8 C for the fresh and aged cells. All the elementary polarizations for the aged cell exceeded those for the fresh cells, strongly indicating that the elementary resistances in the cell increased with cell aging. The proportional contribution of elementary polarization to total polarization (Figure 12(b)), reproduced from the transients, proved that the reaction corresponding to the R_3 element and the solid-state diffusion process significantly contributed to the total polarization during the initial and later stages, respectively, of continuous discharging for 10 s, irrespective of the fresh and aged cells. Furthermore, the power degradation of the cell after aging was mainly ascribed to the increase in the R_3 value of the cell; *i.e.*, the contribution of R_3 increased after cell aging by more than 5%, which is a maximum among the increases in contribution of all the elementary polarizations.

The preceding analysis helps us roughly quantify the contribution of the individual reactions with different time constants; however, a two-electrode electrochemical cell places a serious limitation on the reliable differentiation of the time constants of the real reaction steps. That is, the impedance spectrum obtained from a two-electrode electrochemical cell is significantly distorted from the spectrum of the electrode of concern (*i.e.*, the cathode) due to the overlap of the relaxation times for all the reactions on the anode and cathode sides. The separation in the contributions of the cathode and anode, together with setting the design strategy of the materials, will be discussed in a subsequent section.

3. Analysis Based on a Three-Electrode Electrochemical Cell

Shown in Figure 13 are potential transients during the discharging for 10 s at different rates, obtained from the fresh and aged cells in a three-electrode electrochemical cell configuration. With an increasing number of aging cycles, the transients at the same discharging rate fell faster, indicating that the cell resistance increased, and thereby the power performance of the cell was degraded with cycling. To clarify the reaction steps to most affect the total polarization in the course of the constant-current pulse discharging, and at the same time explore the main factors in power degradation with cell aging, the impedance spectra of the cathode and anode were separately measured at different levels of cell aging.

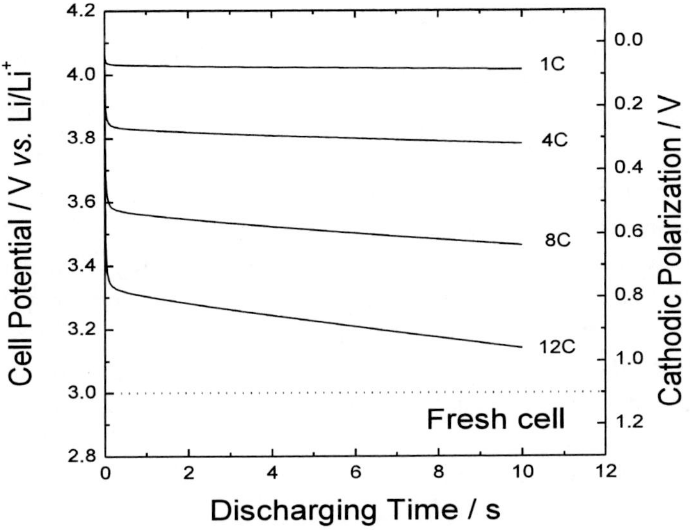

(a)

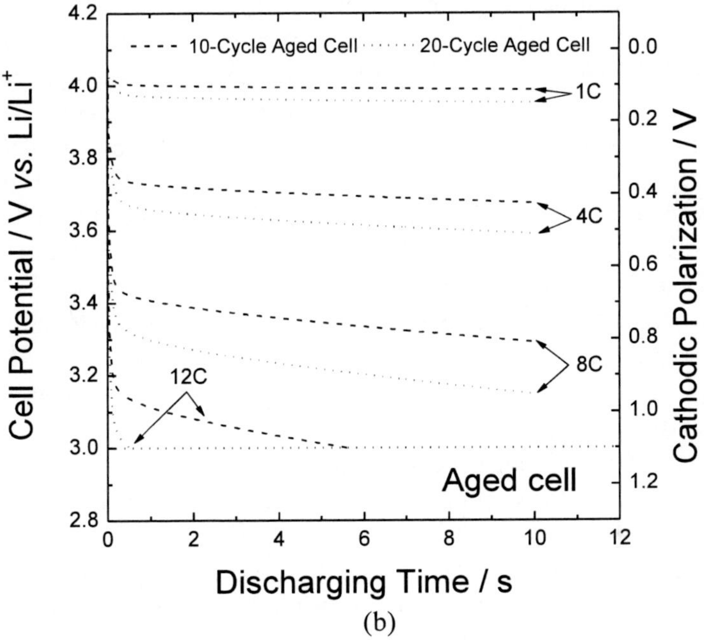

(b)

Figure 13. Experimental cell potential (or cathodic polarization) transients during the cathodic pulse discharging for 10 s at different rates, obtained from (a) the fresh cell and (b) the aged cell.

The impedance spectra of the cathode clearly exhibited two semicircles and an inclined line, as shown in Figure 14(a). These are attributable to the migration/charging of lithium ions in the SEI on the cathode, charge transfer combined with double layer charging at the interface, and lithium diffusion through $LiCoO_2$ oxide, respectively. On the other hand, several semicircles and an intermediate frequency inductive loop were characteristic of the impedance spectra of the lithium anode (Figure 14(b)), the origin of which is yet to be determined. The summed spectra of the cathode and anode, denoted by the open circle in Figure 14(c), revealed a quantitative coincidence with the impedance spectra measured under a two-electrode electrochemical cell configuration. This strongly indicates that the impedance spectra of the cathode and anode were successfully separated using our three-electrode electrochemical cell.

From the impedance spectra of the full cell (Figure 14(c)), it appeared that the first semicircle was nearly invariant, whereas the second semicircle increased with cell aging. Accordingly, a conclusion might be drawn that the SEI resistance of the cathode was almost constant, but its charge transfer resistance was significantly raised with cell aging. However, the separated impedance spectra (Figures. 14(a) and (b)) tell a different story: the SEI resistance of the cathode in the high-frequency region increased with aging, while the high-frequency impedance of the anode decreased. That is, the increase in the SEI resistance of the cathode was offset by the decrease in high-frequency impedance of the anode, making the high-frequency impedance of full cell look invariant. Furthermore, the second semicircles of the full cell impedance spectra of the aged cell reflected the intermediate-frequency inductive loop of the impedance spectra for the anode. Actually, a large inductive loop on the anode impedance spectrum considerably lowered and depressed the second semicircle on the full cell impedance spectra.

Shown in Figures 15(a) and (b) are the equivalent circuits that model the reactions in the cathode and anode, respectively. In particular, the equivalent circuit for the lithium anode included the inductive element parallel to three serial R-CPE elements, which proved to fit the impedance spectra containing an intermediate-frequency inductive loop (although no physical meaning can be attached to this at present [31, 38, 39]). The equivalent circuit for the full cell can be constructed from the combination of the circuits for the cathode and anode (Figure 15(c)).

For a theoretical analysis of the equivalent circuit, the Warburg impedance of Figure 15(a) was replaced with a transmission line and all the CPEs were assumed to be pure capacitors for the sake of simplicity. All of the electrical parameters of the reconstructed equivalent circuits were determined as described in Sections III-2 and IV-2, and are summarized in Table 1. Shown in Figure 16 are the variations of total polarization with discharging time, determined from the theoretical analysis of the reconstructed equivalent circuit. The calculated transients of total polarization bore a strong quantitative resemblance to the experimental transients.

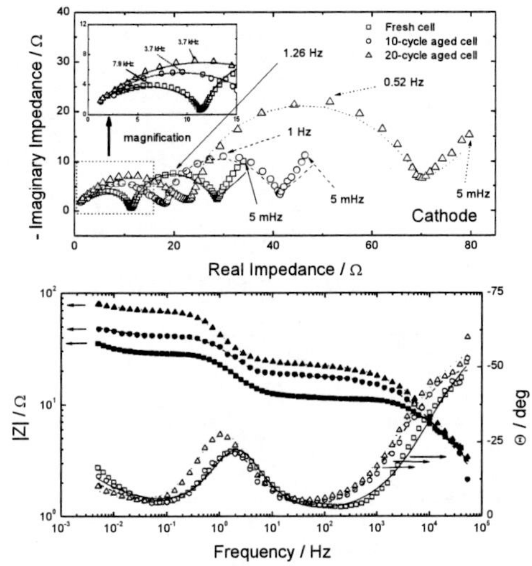

(a) (a-2)

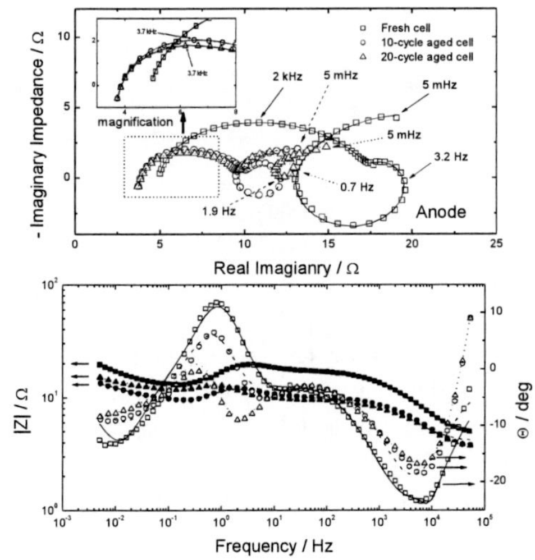

(b) (b-2)

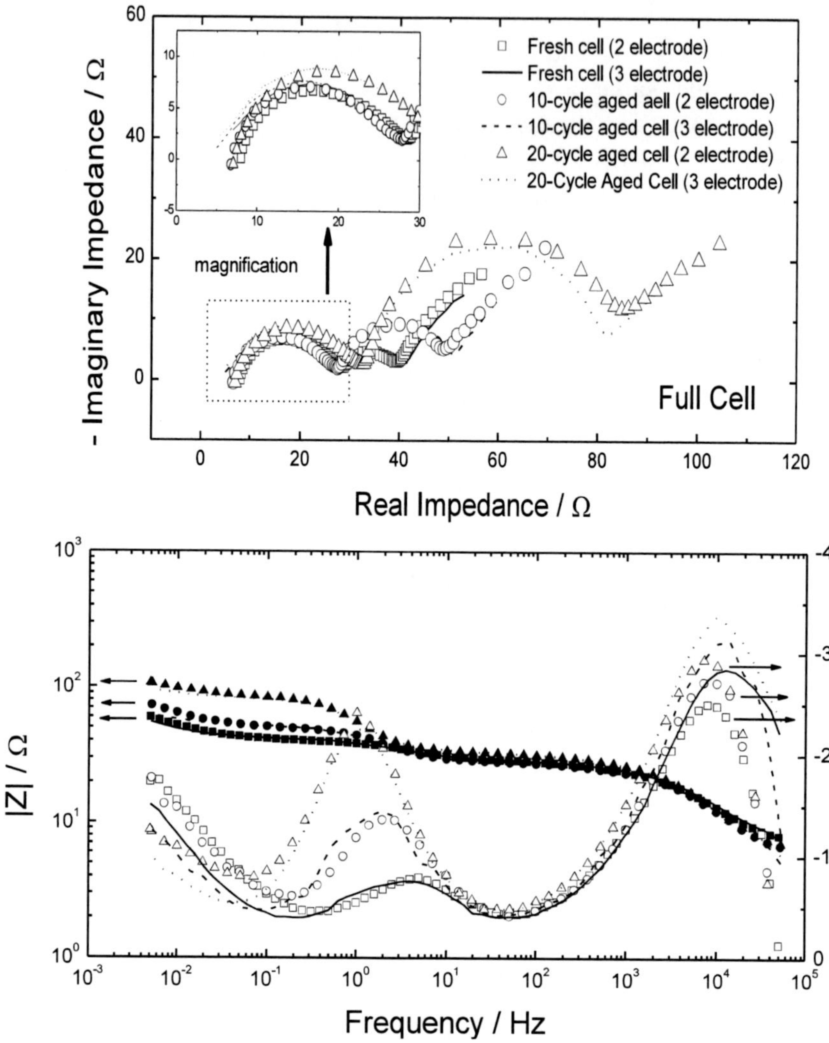

(c) (c-2)

Figure 14. Nyquist plots of (a) cathode, (b) anode, and (c) full cell, obtained from the fresh and aged cells at the cell potential of 4.1 V (*vs.* Li/Li^+). The lines in (a) and (b) were determined from the CNLS fittings of the impedance spectra to the equivalent circuits. In (c), the summation of the impedance spectra of cathode and anode obtained under three-electrode electrochemical cell configuration (lines) were compared with the spectra measured under two-electrode electrochemical cell configuration (symbols). (a)-2, (b)-2, and (c)-2 represent the corresponding Bode plots.

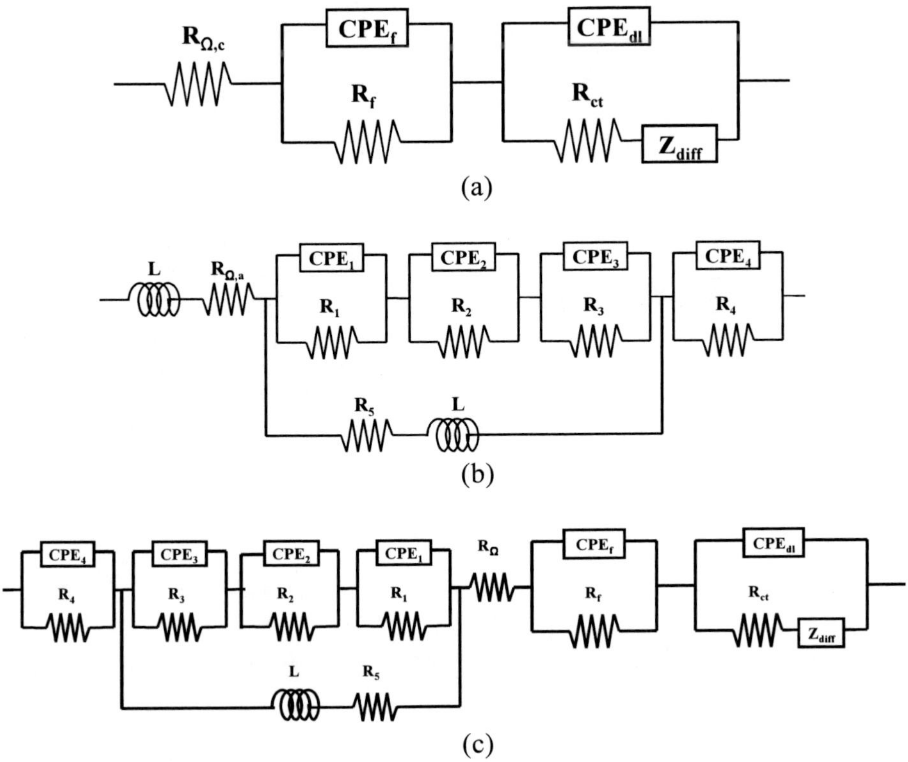

Figure 15. Proposed equivalent circuits to model the reactions in (a) the cathode and (b) the anode. (c) is the equivalent circuit for full cell, obtained from the combination of (a) and (b).

The dependence of elementary polarizations on discharging time for the fresh and aged cells were theoretically separated from the total polarization (Figure 17), and the time-dependent proportional contributions of the elementary polarizations to total polarization were calculated accordingly (Figure 18). The main reason for the power degradation of the aged cell proved to be a significant increase in polarization due to the charge transfer resistance of the cathode; *i.e.*, its proportional contribution to total polarization increased from 30% (for the fresh cell) to 50% (for the 20-cycle aged cell), which was greater than the increases in contribution caused by any other elementary polarizations. This strongly indicates that priority in cell design for the sustained operation of a high-power LIB should be given to surface stabilization of active materials from, *e.g.*, the addition of structure-stabilizing elements [40, 41] and/or surface modification [42, 43].

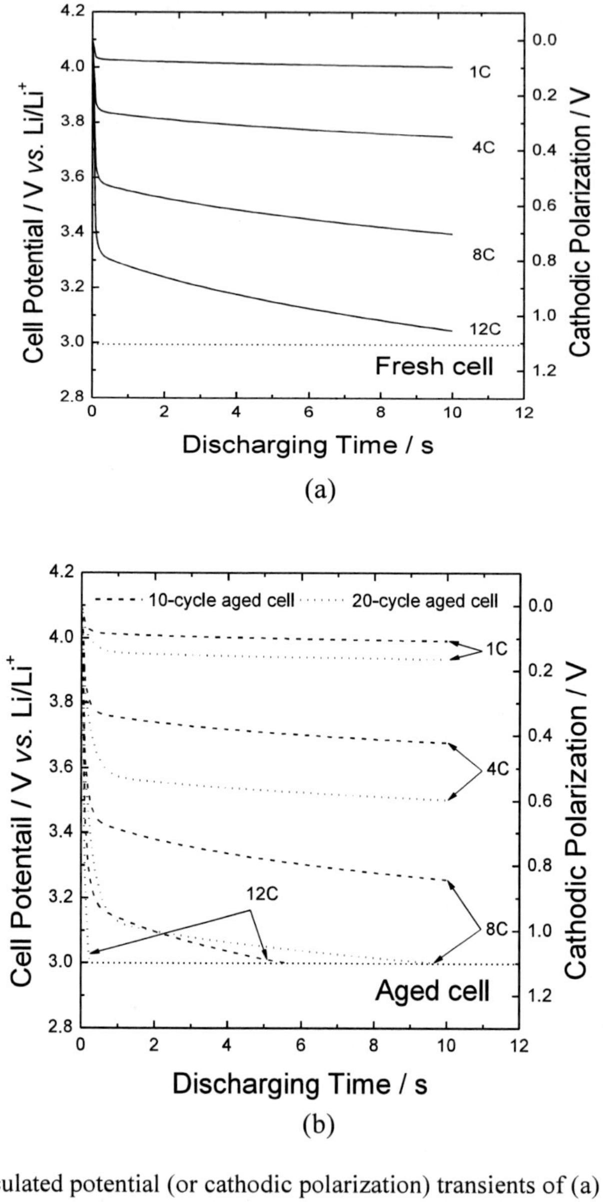

Figure 16. Calculated potential (or cathodic polarization) transients of (a) the fresh cell and (b) the aged cell during the cathodic pulse discharging for 10 s at different rates, determined from the theoretical analysis of the re-constructed equivalent circuit of Figure 15.

Table 1. Electrical parameters of (a) cathode and (b) anode at various levels of aging, determined from the complex non-linear least squares (CNLS) fitting of impedance spectra to the equivalent circuits. (a) includes the chemical diffusion coefficient *D*, diffusion length *L* and some values calculated therefrom.

(a)

	R_Ω (Ω)	R_f (Ω)	CPE_f[2]		R_{ct} (Ω)	CPE_{dl}[2]		A_ω ($\Omega \cdot s^{-0.5}$)	D[3] ($\times 10^{-11} m^2 \cdot s^{-1}$)	L[4] (μm)	R_D[5] (Ω)	C_D[5] (F)	r[5] ($\times 10^{6} \Omega \cdot m^{-1}$)	c[5] ($\times 10^6 s \cdot \Omega^{-1} \cdot m^{-1}$)
			C ($\mu F \cdot s^{\eta-1}$)	η		C ($mF \cdot s^{\eta-1}$)	η							
Fresh	<0.1[1]	11.86	31	0.71	14.95	7.7	0.98	1.92	5.71	5.0	1.27	0.35	0.25	0.07
10-Cycle Aged	"	17.77	42	0.71	20.81	6.4	0.99	2.34	3.87	5.0	1.88	0.35	0.38	0.07
20-Cycle Aged	"	23.31	50	0.68	44.01	5.5	0.96	2.91	3.88	5.0	2.34	0.30	0.47	0.06

(1) Fitted values were close to zero due possibly to the artifact of the cell.

(2) CPE is expressed in the form of $C(j\omega)^\eta$

(3) *D* at 4.1 V (vs. Li/Li^+) was estimated from the galvanostatic intermittent titration technique.

(4) *L* was the radius of the particle, determined from the SEM observation.

(5) $R_D=A_\omega \cdot L \cdot D^{-1/2}$, $C_D=L^2 \cdot R_D^{-1} \cdot D^{-1}$, $r=R_D \cdot L^{-1}$, $c=C_D \cdot L^{-1}$.

(b)

	R_Ω (Ω)	R_1 (Ω)	CPE_1		R_2 (Ω)	CPE_2		R_3 (Ω)	CPE_3		R_4 (Ω)	CPE_4		R_5 (Ω)	L (H)
			C ($\mu F \cdot s^{\eta-1}$)	η		C ($\mu F \cdot s^{\eta-1}$)	η		C ($mF \cdot s^{\eta-1}$)	η		C ($F \cdot s^{\eta-1}$)	η		
Fresh	5.1	3.43	3.95	1	8.57	84.5	0.83	3.31	9.32	0.92	14.82	0.84	0.69	13.41	4.52
10-Cycle Aged	4.2	1.78	0.82	1	3.54	47.4	0.96	10.7	55.6	0.60	8.78	0.90	0.56	6.85	2.09
20-Cycle Aged	4.2	1.67	1.12	1	3.46	83.8	0.93	39.9	51.0	0.71	7.31	2.25	0.70	9.84	3.37

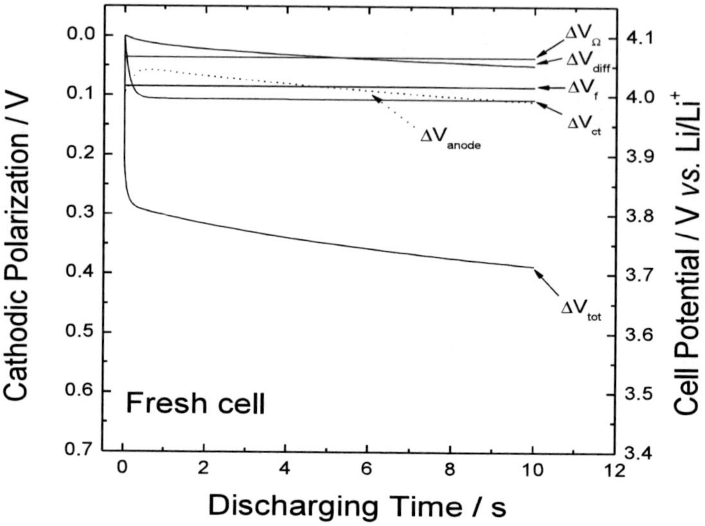

(a)

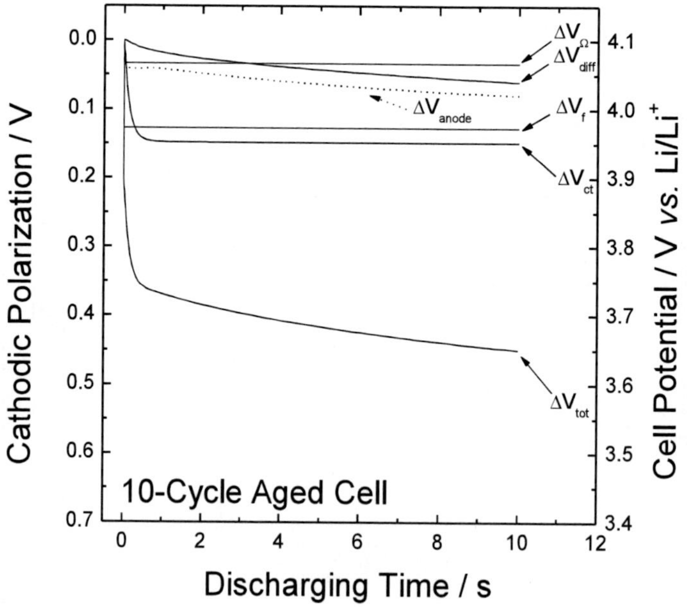

(b)

Figure 17. (*Continued*)

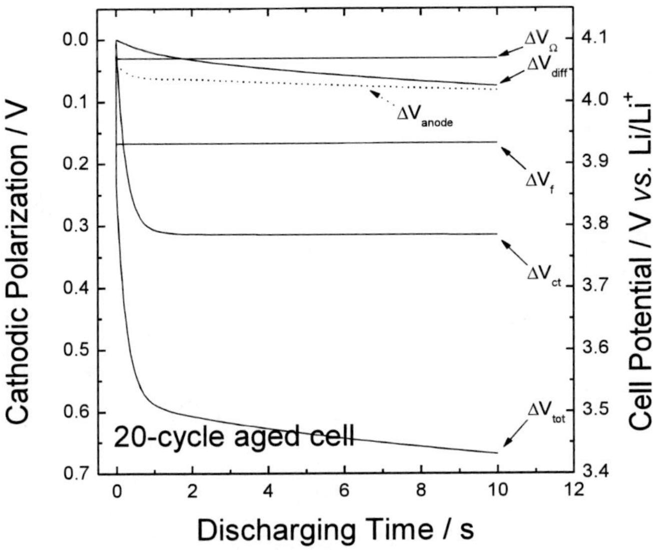

(c)

Figure 17. Variations of elementary and total polarizations, obtained from (a) the fresh cell, (b) the aged cell, with discharging time at 4C rate. ΔV_{anode} is the sum of the contributions of the elementary polarizations in the anode side.

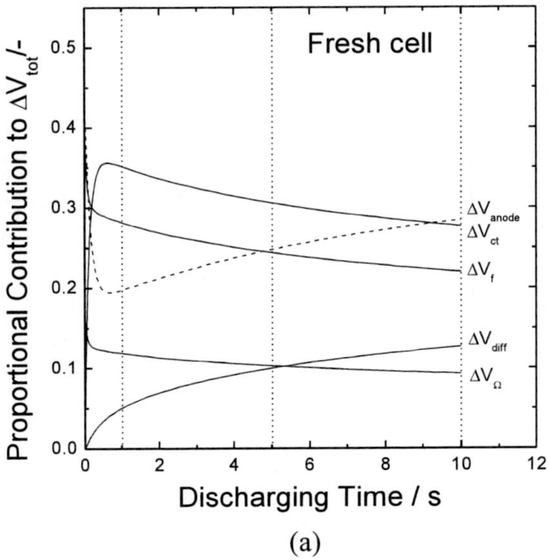

(a)

Figure 18. (*Continued*)

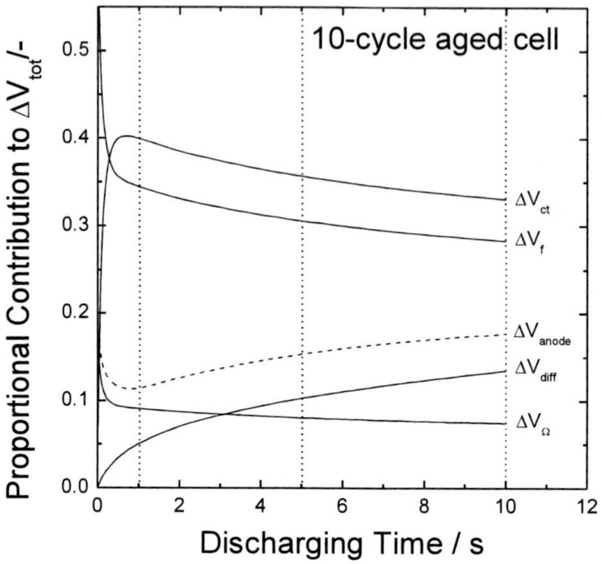

(b)

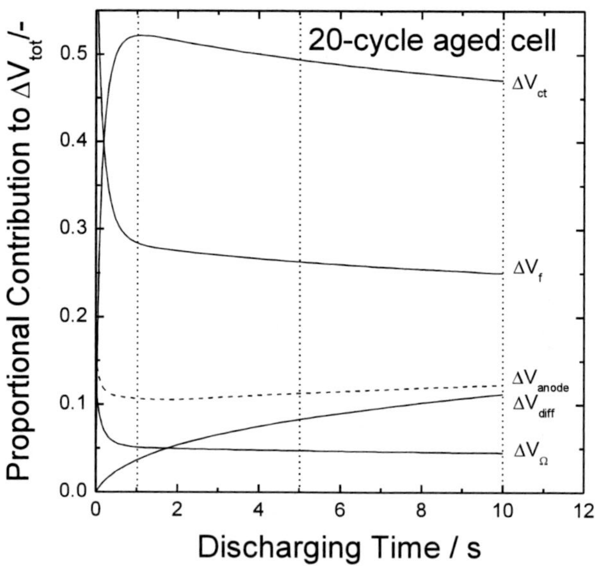

(c)

Figure 18. Time-dependent proportional contribution of elementary impedances to total polarization for (a) the fresh, (b) the 10-cycle aged cell, and (c) the 20-cycle aged cell, reproduced from Figure 17.

V. CRITICAL FACTORS FOR LOW-TEMPERATURE POWER DECLINE

1. Brief Description of Electrochemical Test Procedures

A three-electrode electrochemical cell was used to analyze the effect of temperature on power performance. The components of the cell and its fabrication procedure are the same as described in Section IV-3. For the electrochemical tests, the as-prepared cell was initially cycled several times for activation. Then, the electrochemical impedance spectra at a cell potential of 4.1 V (*vs.* Li/Li^+) were obtained at temperatures ranging from 25 to 5 °C for a frequency range of 50 kHz to 3 mHz. Current pulse tests were carried out immediately after the impedance tests at each temperature by applying different cathodic (discharging) currentsfor 5 s.

2. Effect of Temperature on Total and Elementary Polarizations

Since the reactions associated with battery operation are a thermal activation process, the power density of the battery is significantly reduced with decreasing temperature. The polarization transients at different temperatures manifested the power decline at low operating temperatures, as shown in Figure 19: at lower temperatures, the transients fell faster.

To investigate the main factors in the low-temperature (LT) power decline, the impedance spectra were first determined at different temperatures (Figure 20). In the case of the impedance spectra for the cathode, the first and second semicircles, caused by the reactions in the SEI and the interfacial charge transfer combined with double layer charging, respectively, increased in size with lowering temperature while the inclined line due to the solid-state diffusion process became shorter. From the separated semicircles in the impedance spectra for the anode measured at the low temperatures, it is proved that at least four parallel resistive-capacitive elements were needed to properly model the reactions in the anode during battery operation, the basis for which is not yet clearly understood.

The equivalent circuits of Figure 15 are still valid to model the impedance spectra for the cathode and anode. The diffusion impedance was substituted for the transmission line, and its elementary resistance and capacitance were evaluated in the same manner as in previous sections. All of the electric parameters of the reconstructed equivalent circuits are summarized in Table 2. We note that the higher temperature sensitivity of the charge transfer resistance of the cathode means a larger activation energy for it compared to resistance for Li^+ migration through the SEI, as demonstrated in Figure 21. This further indicates that the low temperature power decline is mainly caused by difficulty in the interfacial charge transfer reaction at low operating temperatures, as is consistent with previous works [12, 44].

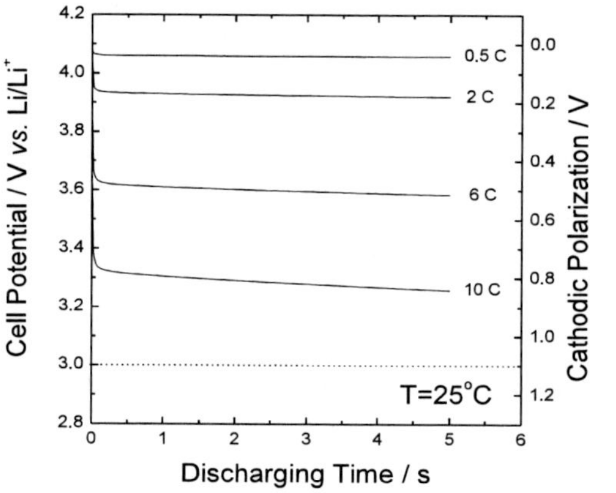

(a)

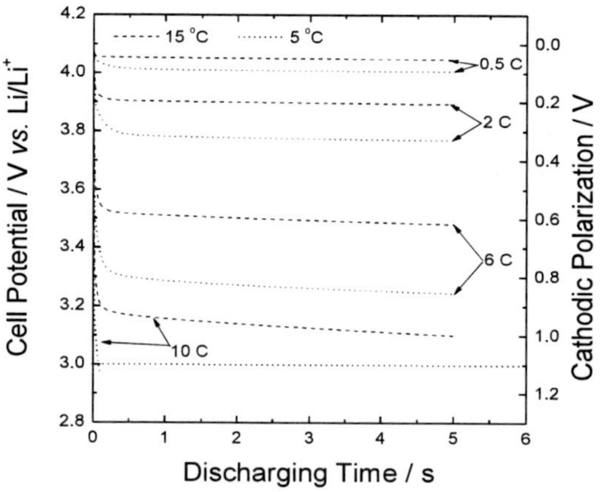

(b)

Figure 19. Experimental cell potential (or cathodic polarization) transients during the cathodic pulse discharging for 5 s at different rates, obtained at the operating temperatures of (a) 25 and (b) 15 and 5 ℃.

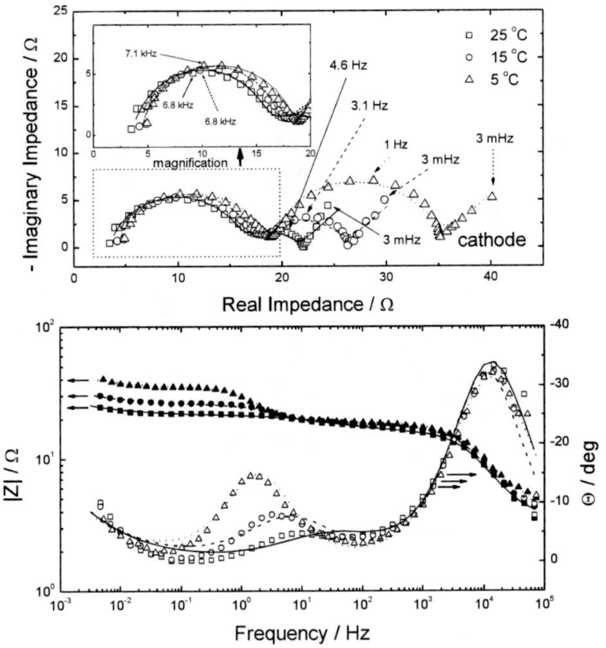
- Imaginary Impedance / Ω
Real Impedance / Ω
25 °C
15 °C
5 °C
7.1 kHz
6.8 kHz
6.8 kHz
magnification
4.6 Hz
3.1 Hz
1 Hz
3 mHz
3 mHz
3 mHz
cathode
|Z| / Ω
Θ / deg
Frequency / Hz

(a) (a-1)

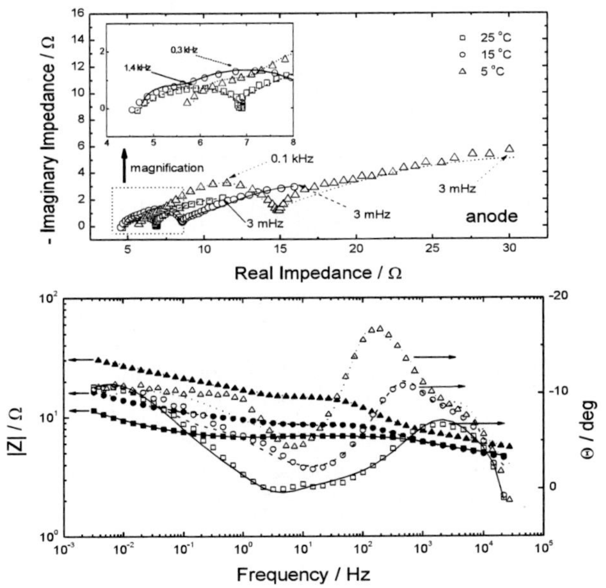
- Imaginary Impedance / Ω
Real Impedance / Ω
25 °C
15 °C
5 °C
0.3 kHz
1.4 kHz
magnification
0.1 kHz
3 mHz
3 mHz
3 mHz
anode
|Z| / Ω
Θ / deg
Frequency / Hz

(b) (b-1)

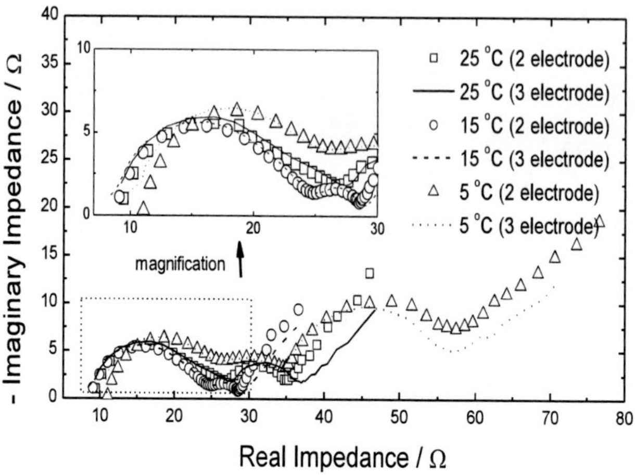

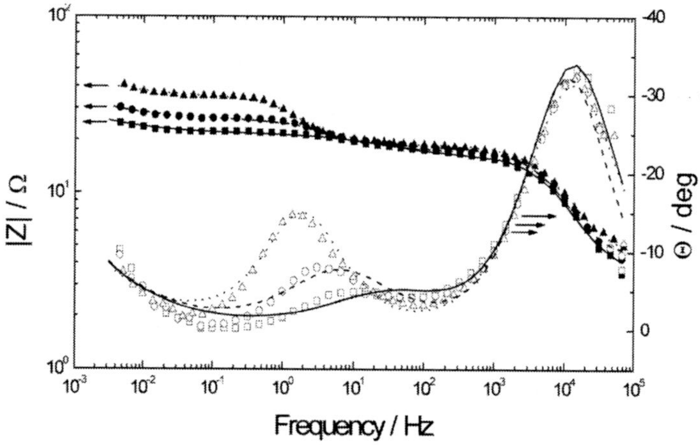

(c) (c-1)

Figure 20. Nyquist plots of (a) cathode, (b) anode, and (c) full cell at the cell potential of 4.1V (*vs*. Li/Li^{+}) and the temperatures of 25, 15, and 5 ℃. The lines in (a) and (b) were determined from the CNLS fittings of the impedance spectra to the equivalent circuits. In (c), the summation of the impedance spectra of cathode and anode obtained under three-electrode electrochemical cell configuration (lines) were compared with the spectra measured under two-electrode electrochemical cell configuration (symbols). (a)-2, (b)-2, and (c)-2 represent the corresponding Bode plots.

Table 2. Electrical parameters of (a) cathode and (b) anode at various operating temperatures, determined from the complex non-linear least squares (CNLS) fitting of impedance spectra to the equivalent circuits. (a) includes the chemical diffusion coefficient *D*, diffusion length *L* and some values calculated therefrom.

(a)

Temp. (°C)	R_Ω (Ω)	R_f (Ω)	CPE_f		R_{ct} (Ω)	CPE_{dl}		A_ω ($\Omega \cdot s^{-0.5}$)	D ($\times 10^{-11} m^2 \cdot s^{-1}$)	L (μm)	R_D (Ω)	C_D (F)	r ($\times 10^6 \Omega \cdot m^{-1}$)	c ($\times 10^6 s \cdot \Omega^{-1} \cdot m^{-1}$)
			C ($\mu F \cdot s^{\eta-1}$)	η		C ($mF \cdot s^{\eta-1}$)	η							
25	3.5	10.57	7.7	0.88	8.48	6.8	0.65	0.76	4.71	5.0	0.55	0.96	0.11	0.19
15	4.0	11.33	6.2	0.91	10.59	8.9	0.74	0.92	4.67	5.0	0.67	0.79	0.13	0.16
5	4.1	12.01	11	0.83	16.66	8.7	0.91	1.13	4.50	5.0	0.84	0.66	0.17	0.13

(b)

Temp. (°C)	R_Ω (Ω)	R_1 (Ω)	CPE_1		R_2 (Ω)	CPE_2		R_3 (Ω)	CPE_3		R_4 (Ω)	CPE_4		R_5 (Ω)	L (H)
			C ($F \cdot s^{\eta-1}$)	η		C ($F \cdot s^{\eta-1}$)	η		C ($mF \cdot s^{\eta-1}$)	η		C ($mF \cdot s^{\eta-1}$)	η		
25	4.2	5.1	5.2	0.88	1.4	1.6	0.93	1.7	0.13	0.77	0.98	0.22	1	N/A	N/A
15	4.0	7.7	3.4	0.79	4.8	0.26	0.54	2.3	0.41	0.67	2.09	0.25	0.99	N/A	N/A
5	4.9	6.3	0.3m	0.94	28	0.31	0.5	3.3	0.73	0.59	4.28	89.3	0.88	N/A	N/A

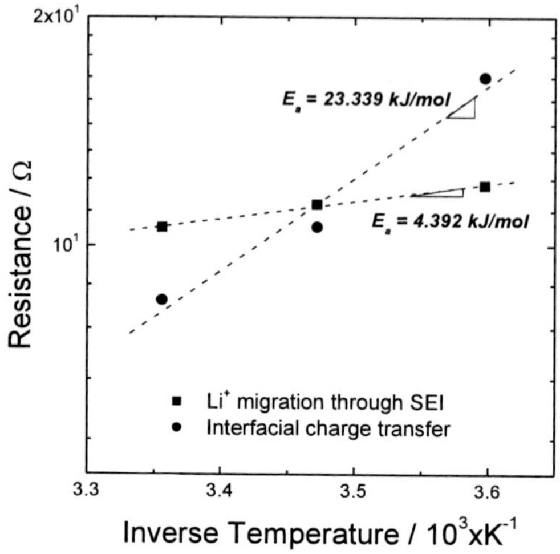

Figure 21. Arrhenius plot of the resistance *vs.* inverse temperature, to determine the activation energies for Li^+ migration through SEI layer and interfacial charge transfer.

The remaining question is the extent to which the charge transfer resistance affects total polarization in the course of battery discharging. Our theoretical analysis of the reconstructed circuits led to the dependence of the elementary polarization transients on operating temperatures (Figure 22), and the proportional contribution of elementary impedances to total polarization were, accordingly, estimated as a function of discharging time (Figure 23). At the operating temperature of 5 °C, the proportional contribution of the interfacial charge transfer resistance to total dc polarization was up to more than 30% at the initial stage of discharging and it reached about 22% even at the later stage of discharging, as shown in Figure 23(c).

3. Power Performance of Hybrid Electrodes

Since the LT power decline of a LIB is intrinsic and unavoidable, hybridization of the typical battery working concept with the capacitor working concept has recently been studied in an attempt to dramatically increase the LT performance of LIBs, due primarily to the superior LT performance of the capacitor [45]. The combination of insertion materials and electrochemical double layer capacitor (EDLC) materials can be the case in point of hybrid electrode. The methodology suggested in this section might be successfully utilized to gain an insight into LT pulse discharging of the cell with hybrid electrode(s). We now consider a hypothetical equivalent circuit for a cell of one hybrid electrode (Figure 24). We note that the electric element for the EDLC materials, *i.e.*, the capacitor is connected in parallel with those elements for the insertion materials, because EDLC materials contribute to charge storing independent of the insertion materials.

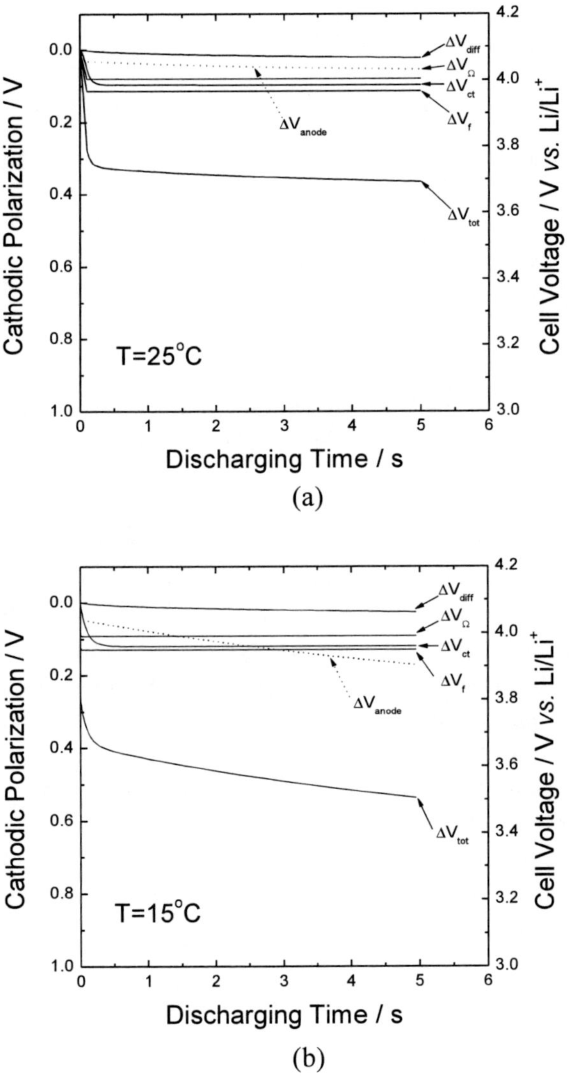

Figure 22. (*Continued*)

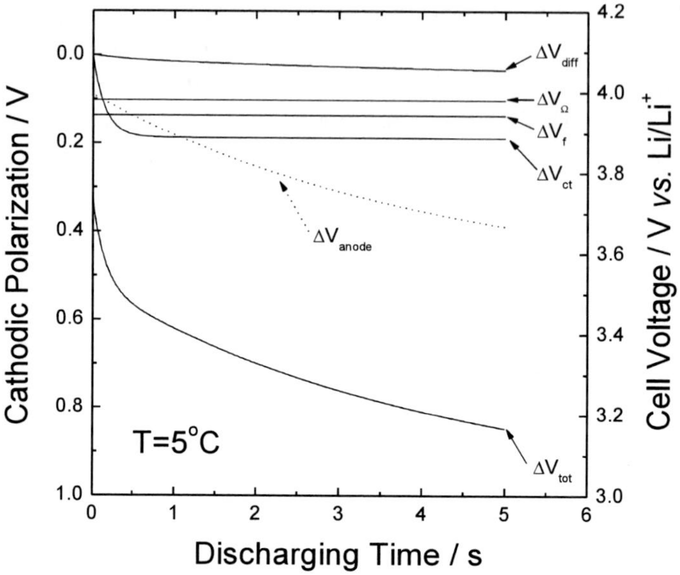

(c)

Figure 22. Variations of elementary and total polarizations with discharging time at 4C rate and at the temperatures of (a) 25, (b) 15 and (c) 5 ℃.

The calculated impedance spectra at different capacitances of the EDLC materials are shown in Figure 25. As EDLC capacitance increases, the real impedance that essentially determines the value of dc polarization in the course of pulse charging/discharging decreases (see the real impedances at 1 Hz, as indicated in the figures). This strongly indicates that the addition of the EDLC materials to the electrode is beneficial in reducing cell impedance.

Potential or polarization transients of the cell with a hybrid electrode differ significantly from transients of a cell consisting only of typical insertion electrodes: Figures 26(a) and (b) show the variations of potential or dc polarization with discharging time for various EDLC capacitance values, calculated for the nodes of $V_{hybrid,L}/V_{hybrid,R}$ (for a hybrid electrode) and $V_{tot,L}/V_{tot,R}$ (for a full cell), respectively, of Figure 24. It is noted that the capacitance of the EDLC materials has a strong influence on the polarization due to the hybrid electrode (Figure 26(a)). That is, a milder decrease in total dc polarization with larger EDLC capacitance (Figure 26(b)) is totally ascribed to reduced polarization of the hybrid electrode.

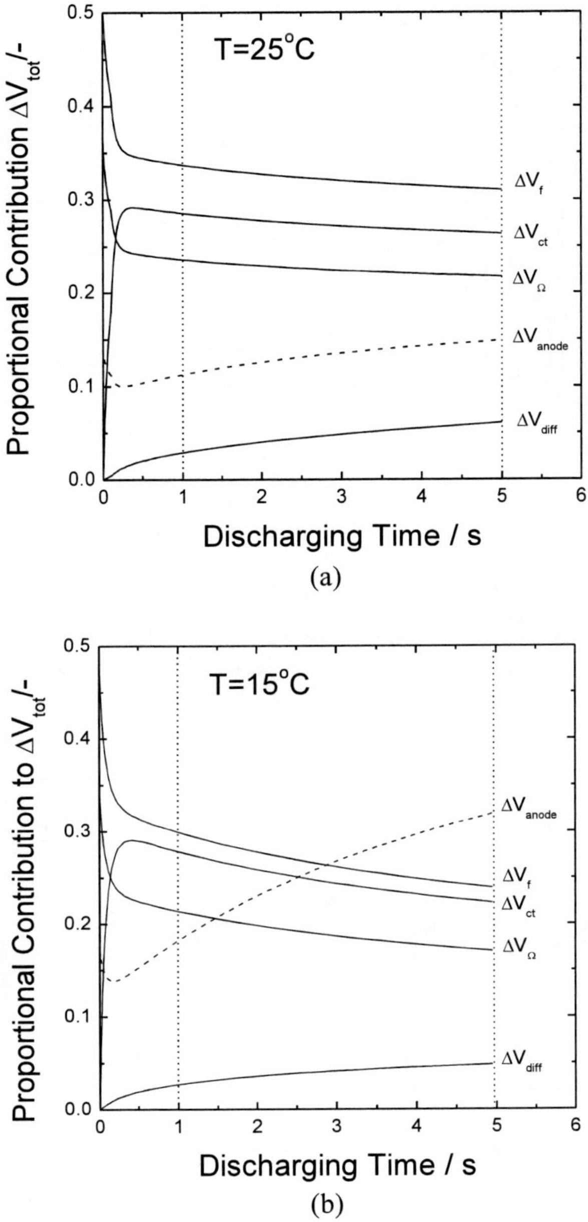

Figure 23. (*Continued*)

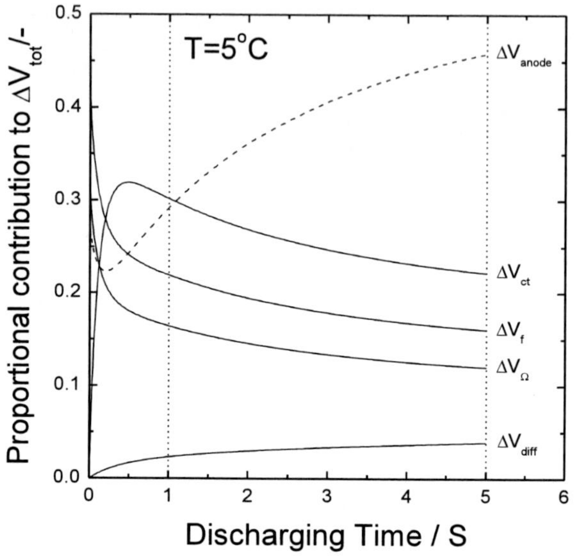

(c)

Figure 23. Time-dependent proportional contributions of elementary impedances to total polarization at the temperature of (a) 25, (b) 15 and (c) 5 ℃, reproduced from Figure 22.

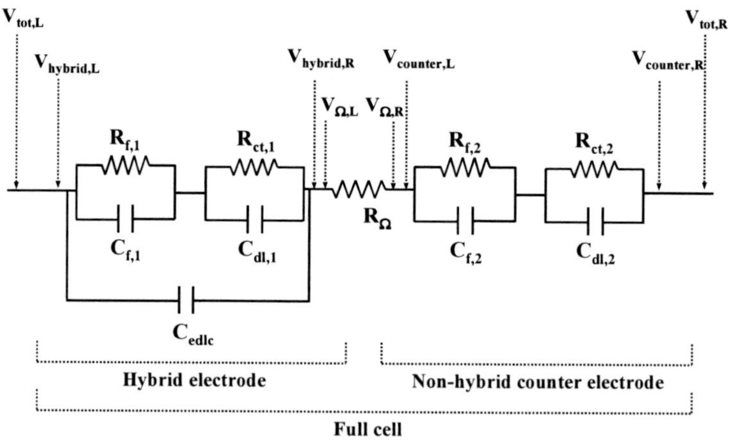

Figure 24. Hypothetic equivalent circuit for a cell of one hybrid electrode where the insertion materials and electrochemical double layer capacitor (EDLC) materials are conceptually combined.

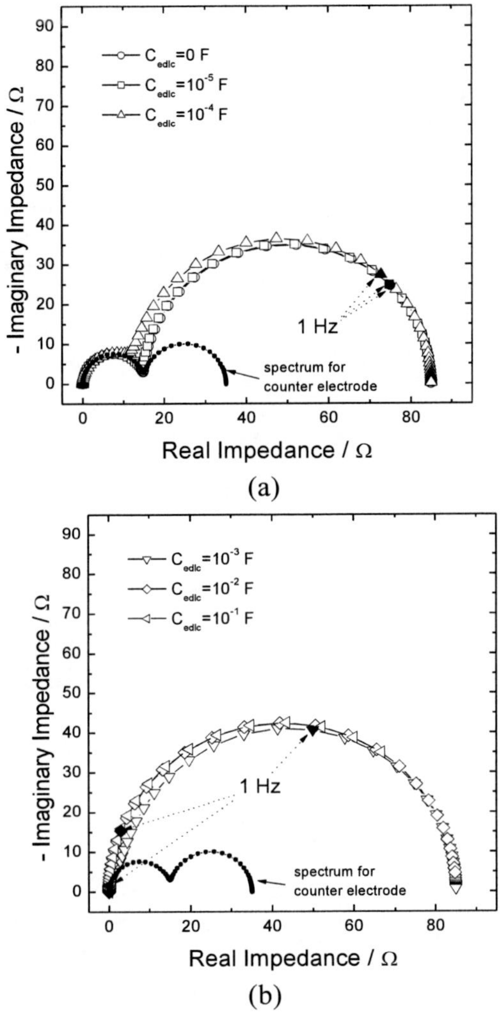

Figure 25. Calculated impedance spectra at different capacitances of the EDLC materials. The closed symbols indicate the impedance values at 1 Hz.

When EDLC capacitance is greater than 1 F in this hypothetical system, there is virtually no contribution of the hybrid electrode to total dc polarization during the continuous discharging for 5 s, while total polarization depends entirely on polarization due to the non-hybrid counter electrode. Accordingly, it is anticipated that LT power decline can be greatly reduced when the insertion materials, whose reaction impedance is highly temperature-sensitive, become hybridized with the EDLC materials.

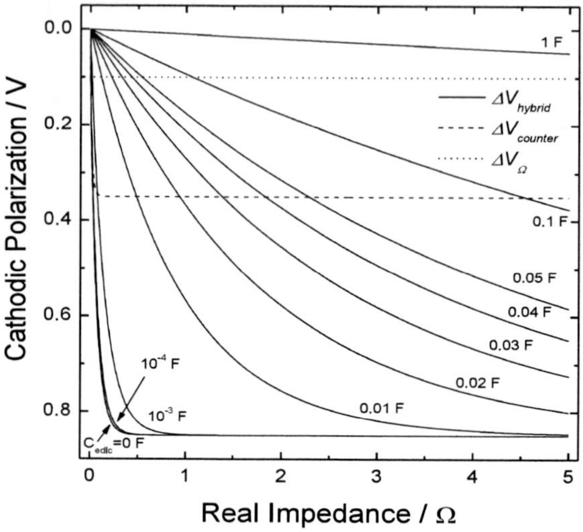

(a)

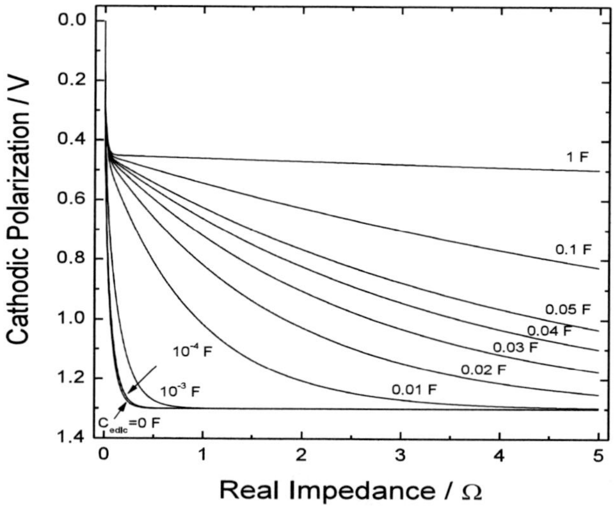

(b)

Figure 26. Variations of potential (or cathodic polarization) transients during the cathodic pulse discharging for 5 s at various EDLC capacitance values, calculated for (a) hybrid electrode and (b) full cell. Potential transients obtained for a non-hybrid counter electrode ($\Delta V_{counter}$) and an electrolyte (ΔV_{Ω}) were included in (a) for the sake of comparison.

VI. CONCLUSION

The successful design of a high-power LIB is critically dependent on the extent to which the effect of cell components on cell polarization is quantitatively understood during cell operation. This study suggests a viable way to quantify the time-dependent proportional contribution of elementary polarizations (or individual reaction steps) during the pulse discharging process, and provides a good starting point for high power cell design. The present work can be summarized as follows:

1. Dependence of the contribution of elementary impedances to total dc polarization on battery discharging time was successfully analyzed based on a combination of electrochemical impedance spectroscopy and theoretical analysis of equivalent circuit.
2. Interfacial charge transfer resistance of the cathode proved to be the most important resistance factor in the course of high-rate battery discharging at ambient and low operating temperatures. Improvement of the electrochemical activity of the cathode is the appropriate way to effectively lower the total dc polarization during battery operation and enhance battery power. A hybrid electrode for which battery and capacitor working concepts are combined might significantly elevate the power density, particularly for low temperature operation.
3. From our comparative analysis on the time-dependent contribution of elementary polarizations of fresh and aged cells, power degradation after repeated discharge-charge cycles were attributable to a significant increase in polarization due to charge transfer resistance of the cathode. Stabilization of the surface structure of the active materials would have priority over any other design strategies in order to secure sustained high-power operation of a LIB.
4. Much caution should be taken regarding the effect of the lithium anode on impedance spectra and polarization transients when they are obtained under a two-electrode electrochemical cell configuration. The impedance and polarization caused by a lithium anode significantly changes the overall spectra and transients, respectively, in value and shape.
5. The method suggested in this study gives us a viable way to determine the critical factors for battery power, thereby enabling us to systematically design a high-power LIB. Furthermore, this approach could possibly be applied to the diagnosis of a battery with degraded

power, and gives insight into the design of a high-power battery with long shelf and calendar life.

6. The successful application of our method requires the construction of equivalent circuits to perfectly model the cell reactions and at the same time needs the information on the correlation between elementary resistances and design parameters of cell components. Nevertheless, an in-depth investigation on the discrepancy, after the redesign of the cell, between calculated (or expected) power performance and experimental performance, might make our understanding of the reaction mechanism and the core design parameters more complete. This leads to the virtuous cycle of mechanism elucidation, components design, cell preparation, and diagnosis of high-power LIBs.

ACKNOWLEDGMENTS

This work was supported by a Korea Research Foundation Grant funded by the Korean Government (MOEHRD) (KRF-2006-331-D00713). Furthermore, this work was partially supported by a grant-in-aid for the National Core Research Centre Program from MOST and KOSEF (No. R15-2006-022-01001-0).

REFERENCES

[1] Horiba, T; Maeshima, T; Matsumura, T; Koseki, M; Arai, J; Muranaka, Y. *Journal of Power Sources.,* 2005, 146 107.
[2] Sawai, K; Yamato, R; Ohzuku, T; *Electrochim. Acta.,* 2006, 51, 1651.
[3] Chen, ZH; Amine, K. *J. Electrochem. Soc.*, 2006, 153, A1221.
[4] Kardena, E; Shinnb, P; Bostockc, P; Cunninghamc, J; Schoultzd, E; Koka, D. *Journal of Power Sources.,* 2005, 144, 505.
[5] Karden, E; Ploumen, S; Fricke, B; Miller, T; Snyder, K. *Journal of Power Sources.,* 2007, 168, 2.
[6] Yoskiike, N; Ayusawa, M; Kondo, S. *J. Electrochem. Soc.*, 1984, 131, 2600.
[7] Thomas, MGSR; Bruce, PG; Goodenough, JB. *J. Electrochem. Soc.*, 1985, 132, 1521.
[8] Levi, MD; Salitra, G; Markovsky, B; Teller, H; Aurbach, D; Heider, U;

Heider, L. *J. Electrochem. Soc.*, 1999, 146, 1279.
[9] Abraham, KM; Pasquariello, DM; Willstaedt, EM. *J. Electrochem. Soc.*, 1998, 145, 482.
[10] Zhang, S.S. *Journal of Power Sources.,* 2007, 164, 351.
[11] Aurbach, D. *Journal of Power Sources.,* 2000, 89, 206.
[12] Zhang, XW; Wang, C; Appleby, AJ. *Journal of Power Sources* 2003, 114, 121.
[13] Shiao, HC; Chua, D; Lin, HP; Slane, S; Salomon, M. *Journal of Power Sources.,* 2000, 87, 167.
[14] Lin, HP; Chua, D; Salomon, M; Shiao, HC; Hendrickson, M; Plichta E; Slane, S. *Electrochem. Solid-State Lett.*, 2001, 4, A71.
[15] Frackowiak, E; Gautier, S; Gaucher, H; Bonnamy, S; Beguin, F. Carbon 1999, 37, 61.
[16] Yang, ZH; Wu, HQ. *Chem. Phys. Lett.*, 2001, 343, 235.
[17] Li, H; Shi, L; Wang, Q; Chen, L; Huang, X. *Solid State Ionics.,* 2002, 148, 247.
[18] Kang, K; Ceder, G. *Phys. Rev. B.,* 2006, 74, 094105.
[19] Kang, K; Meng, YS; Breger, J; Grey, CP; Ceder, G. *Science.,* 2006, 311, 977.
[20] Xie, J; Kohno, K; Matsumura, T; Imanishi, N; Hirano, A; Takeda, Y; Yamamoto, O. *Electrochim. Acta.*, 2008, 54, 376.
[21] Boukamp, BA. *Solid State Ionics.,* 1986, 20, 31.
[22] Macdonald, JR; Potter, LD. *Solid State Ionics.,*1987, 24, 61.
[23] Franklin, AD; de Bruin, HJ. *Phys. Stat. Sol.*, 1983, 75, 647.
[24] Smirnova, AL; Ellwood, KR; Crosbie, GM. *J. Electrochem. Soc.*, 2001, 148, A610.
[25] Schichlein, H; Müller, AC; Voigts, M; Krügel, A; Ivers-Tiffee, E. *J. Appl. Electrochem.*, 2002, 32, 875.
[26] Kim, JS; Pyun, SI; Shin, HC; Kang, SJL. *J. Electrochem. Soc.*, 2008, 155, B762.
[27] Stoynov, Z; Savova-Stoynov, B. *J. Electroanal. Chem.*, 1986, 209, 11.
[28] Stoynov, Z. *Electrochim. Acta.,* 1989, 34, 1187.
[29] Stoynov, Z. *Electrochim. Acta.*, 1990, 35, 1493.
[30] Shin, HC; Pyun, SI; Vayenas, CG; Conway, BE; White, RE. (Eds.) Modern Aspects of Electrochemistry, Vol. 36, Kluwer Academic Publishers/Plenum Publishers, NY, 2003, 255.
[31] Cho, HM; Park, YJ; Shin, HC. *J. Electrochem. Soc.,* 2010, 157, A8.
[32] Kang, DK; Shin, HC. *J. Solid State Electrochem.*, 2007, 11, 1405.
[33] Barsoukov, E; Macdonald, JR. Impedance Spectroscopy, 2nd (Eds.) John Wiley & Sons, Inc., Hoboken, New Jersey, 2005.
[34] Bisquert, J; Garcia-Belmonte, G; Bueno, P; Longo, E; Bulhões, LOS. *J. Electroanal. Chem.*, 1998, 452, 229.
[35] Bisquert, J; Garcia-Belmonte, G; Fabregat-Santiago, F; Bueno, PR. *J.*

Electroanal. Chem., 1999, 475, 152.
[36] Blyr, A; Sigala, C; Amatucci, G; Guyomard, D; Chabre, Y; Tarascon, JM. *J. Electrochem. Soc*., 1998, 145, 194.
[37] Dolle, M; Orsini, F; Gozdz, AS; Tarascon, JM. *J. Electrochem. Soc*., 2001, 148, A851.
[38] Harrington, DA; Conway, BE. *Electrochim. Acta*., 1987, 32, 1703.
[39] Gabrielli, C; Mocoteguy, P; Perrot, H; Nieto-Sanz, D; Zdunek, A. *J. Appl. Electrochem*., 2008, 38, 457.
[40] Abraham, DP; Twesten, RD; Balasubramanian, M; Petrov, I; McBreen, J; Amine, K. *Electrochem. Commun*., 2002, 4, 620.
[41] Blooma, I; Jones, SA; Battaglia, VS; Henriksen, GL; Christophersen, JP; Wright, RB; Hob, CD; Belt, JR; Motloch, CG. *J. Power Sources.,* 2003, 124, 538.
[42] Takamura, T; Eguchi, S; Suzuki, J; Omae, O; Sekine, K. *J. Power Sources.,* 2005, 146, 129.
[43] Lee, JG; Kim, TG; Park, B. *Mater. Res. Bull*., 2007, 42, 1201.
[44] Go, J; Kim, J; Kim, H; Choi, Y; Kim, K; Abstract 747, The Electrochemical Society Meeting Abstracts, Vol. 2007-02, Washington DC, Oct. 7-12, 2007.
[45] Plitz, I; Dupasquier, A; Badway, F; Gural, J; Pereira, N; Gmitter, A; Amatucci, GG. *Appl. Phys. A.,* 2006, 82, 615.

In: Lithium Batteries: Research, Technology… ISBN: 978-1-60741-722-4
Editors: G. R. Dahlin and K. E. Strom

Chapter 4

CHEMICAL OVERCHARGE PROTECTION OF LITHIUM-ION CELLS

Zonghai Chen**, *Yan Qin and Khalil Amine
Chemical Sciences and Engineering Division, Argonne National Laboratory
9700 South Cass Avenue, Argonne, IL 60439, USA

ABSTRACT

Overcharge protection is not only critical for preventing the thermal runaway of lithium-ion batteries, but also important for automatic capacity balancing. This chapter compares three overcharge protection strategies—external circuit protection, inactivation agents, and redox shuttles—to highlight the advantage of redox shuttles for overcharge protection. Then the redox shuttle history and mechanism are introduced and the latest advances on redox shuttles are described. Fundamental studies for designing stable redox shuttles for use in lithium-ion batteries are also discussed.

* Corresponding author: E-mail: Zonghai.chen@anl.gov

INTRODUCTION

Lithium-ion batteries have been widely used to power portable electronic devices and they also have demonstrated promise for large-scale applications, such as hybrid electric vehicles (HEV) [1] and stationary energy backup systems [2, 3]. Thus, the intrinsic energy storage characteristics of Li-ion batteries have attracted significant attention from both academic and industrial communities. When a lithium-ion cell is fully charged, the positive electrode contains a strong oxidizing transition metal oxide (i.e. $Li_{1-x}MO_2$, M=Ni, Co, Mn), while the negative electrode contains lithiated carbon, a very strong reducing material. Sandwiched between the positive electrode and the negative electrode is a non-aqueous electrolyte that uses an organic carbonate solvent and a lithium salt. In the cell, this solvent tends to be readily oxidized and reduced. Thus, the lithium-ion cell itself is thermodynamically unstable and the compatibility of the cell components is kinetically achieved with the presence of the surface passivation films on the electrode surface. Therefore, lithium-ion batteries are very sensitive to thermal and overcharge abuse and pose significant fire hazards. Overcharge of lithium-ion cells can lead to chemical and electrochemical reactions in battery components [4, 5], gas release [4-6], and rapid increase of cell temperature [4-6]. It can also trigger self-accelerating reactions in the batteries, which can lead to thermal runaway and possible explosion [7].

Overcharge generally occurs during the charge of a battery pack with multiple lithium-ion cells connected in series as shown in Figure 1. When a battery pack is charged, the charger generally continuously monitors the voltage of the battery pack to roughly estimate the state of charge (SOC) of the battery pack; it does not monitor each cell and so assumes that each cell in the pack is identical to the others in terms of capacity and SOC. However, this assumption is difficult to validate in real operation for the following reasons:

- It is difficult for the manufacturers to assemble truly identical cells for a specific battery pack; a small variation in cell capacity and cell performance is normal.
- Some of the cells in the pack might age faster than the others, resulting in reduced cell capacity.
- Accidents, such as abuses and mechanical damage, can increase the variation among the cells in the battery pack.
- Finally, after a certain period of operation, the cells in the battery pack can have different SOC from cell to cell.

Therefore, there is always chance that one or more cells in the battery pack has less capacity than the others as shown in Figure 1a, and the cell with less capacity is called weak cell (the second cell in Figure 1a). When this battery pack is charged, the weak cell will reach its top SOC first while the others are still not fully charged (see Figure 1b). At this point, the voltage of the whole pack is still lower than the expected value and the charger will continue to charge the pack and overcharge the weak cell. Therefore, the overcharge protection must operate at the cell level to assure safe operation of the battery pack. The cell-level overcharge protection can also reduce the need for costly cell capacity balancing during battery manufacturing, maintenance, and repair. With the cell-level overcharge protection mechanism, the battery pack can be charged as a whole. When a cell reaches its top of SOC during charge, the cell voltage can electrically trigger the overcharge protection mechanism, and the excess current will be handled by the incorporated overcharge protection mechanism without causing overcharge to the cell. Under this mechanism, the charging process of the battery pack can continue until all the cells are fully charged.

COMPARISON OF AVAILABLE TECHNOLOGIES

Currently, the technologies available for cell-level overcharge protection include external voltage regulation, inactivation agents, and redox shuttles. A brief comparison of these three mechanisms is given in Table 1.

Table 1. Comparison of different overcharge protection mechanisms.

Characteristic	External voltage regulation	Inactivation agent	Redox shuttle
Mechanism	Electronic regulation Reversible	Permanent inactivation of the overcharged cells by gassing and coating of insulator coating. Irreversible.	Active and reversible
Physical device	Electric circuit	Electrolyte Additive	Electrolyte additive
Weight, volume, and cost	Disadvantage	Advantage	Advantage
Heat generation	Electric work	Heat from electrochemical and chemical reactions of the dditive.	Electric work
Thermal management	Advantage	Disadvantage	Disadvantage
Exemplary technologies	Ref. [8, 9]	Ref. [10-12]	Ref. [13-17]

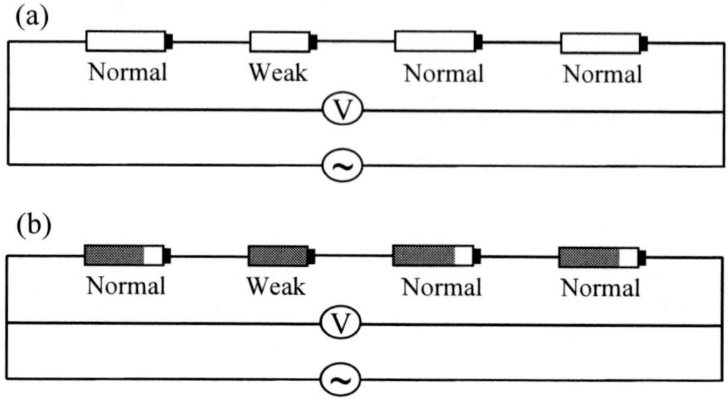

(a) Fully discharged battery packed; (b) partially charged battery pack with the weak cell being fully charged. The blue bar indicates the state of charge of each cell.

Figure 1. Schematic of a battery pack with several cells connected in series on charging.

External voltage regulation [8, 9] is the dominant technology for state-of-the-art lithium-ion batteries. With this technology, the voltage of each lithium-ion cell is continuously monitored by an external circuit board. Once the cell voltage exceeds the regulated value, the external bypassing circuit is activated and the charging current will flow through the external circuit instead of the lithium-ion cell. There are several advantages associated with the external voltage regulation. First of all, the regulated value of circuit board can be fine-tuned according to the specific chemistry of the lithium-ion cells being protected. Hence, external voltage regulation can be a universal technology to provide overcharge protection for a wide range of energy storage devices. In addition, if the overcharge situation occurs to a given cell, the excess current passes through the external circuit instead of the cell, and, the heat converted from the electric current is generated on the external circuit board instead of the protected cell. This can ease the thermal management of the battery system. The disadvantage of external voltage regulation is that it adds complexity, weight, volume, and cost to the battery management system, which can be a major drawback when the energy or power density of the battery is highly demanded.

An alternative overcharge protection mechanism is to use inactivation agents such as 3-thiophenylpropane [11], biphenyl [10] and 3-chlorothiophene [12], and furan [12]. The advantage of this technology is that only a small amount of the inactivation agents will be added to the non-aqueous electrolytes as additives. Hence, there will be only small or no increase in the weight and volume of the battery pack. Overcharge of the cell will result in the polymerization of the additive [10, 12] or an electrochemical reaction that generates a large amount of gas [11]. The resulted insulating polymer will cover on the electrode surface and eventually block the pathway of lithium ions; placing the cell in an open-circuit status. The generated gas can significantly increase the internal pressure of the lithium-ion cell, to a point where it activates the venting device built in the cell and permanently inactivates the cell. Apparently, overcharge of lithium-ion cells incorporating inactivation agents will permanently disable the cell and lead to the failure of the whole battery pack.

The redox shuttle [13-15, 17] is an electrolyte additive that can be reversibly oxidized/reduced at a characteristic potential and provides an intrinsic overcharge protection for lithium-ion batteries that neither increases the complexity and weight of control circuitry nor permanently disables the cell when activated. The redox shuttle molecule (S) has its defined redox potential, at which it can be oxidized on the positive electrode and form a radical cation ($S^{\bullet+}$) (see Equation 1).

$$S \rightarrow S^{\bullet+} + e^- \tag{1}$$

The radical cation then travels to the negative electrode through the electrolyte and is reduced in accordance with Equation 2.

$$S^{\bullet+} + e^- \rightarrow S \tag{2}$$

The redox shuttle molecule then diffuses back to the positive electrode for the next redox cycle, and the electrons move from the positive electrode to the negative electrode through the external circuit. During normal operation, the redox potential of the redox shuttle is not reached and the molecules stay inactive. When the cell is overcharged, the potential of the positive electrode increases, and the redox cycle of the redox shuttle molecules is activated. The net reaction of the redox cycle is to shuttle the charge forced by the external circuit through the lithium-ion cell without also forcing intercalation/ deintercalation of lithium in the electrodes of the cell.

According to above comparison, we can see that redox shuttles have promising applications for high-energy and high-power lithium batteries, particularly for transportation applications, to reduce the weight and volume of the battery pack. It is the major focus of this chapter to discuss the stability and application of redox shuttles for 4V class lithium-ion batteries.

Table 2. Exemplary redox shuttles for overcharge protection of 3V class lithium-ion batteries.

Redox shuttle	Citation	Redox potential V vs. Li^+/Li	Oxidation reaction
I^-	[18, 19]	3.2	$3I^- - 2e \rightarrow I_3^-$
Br^-	[20]	3.78	$3Br^- - 2e \rightarrow Br_3^-$
Ferrocene	[21]	3.05-3.38	$Fc - e \rightarrow Fc^+$ (Fc = ferrocene and its derivatives)
	[22]	3.25	
n-butylferrocene	[21]	3.18-3.50	
	[22]	3.18	
N,N-Dimethylaminomethylferrocene	[21]	3.13-3.68	
1,1-Dimethylferrocene	[21, 23]	3.06-3.34	
1-Acetylferrocene	[22]	3.509	
1-Benzoylferrocene	[22]	3.51	
Methyl Ferrocenecarboxylate	[22]	3.505	
Ferrocenecarboxamine	[22]	3.486	
1-(Dimethylamino)methylferrocene	[22]	3.435	
Ferrocenemethanol	[22]	3.258	
Ferrocenecarboxaldehyde	[22]	3.541	
N,N-Dimethylferrocene	[22]	3.128	

HISTORICAL REVIEW

The research on redox shuttles for overcharge protection of lithium-ion batteries can be traced back to the 1980s, when Behl et al. reported that I^- has its first oxidation potential at about 3.25 V vs. Li^+/Li and hence is suitable for overcharge protection of 3V class lithium-ion batteries [18-19]. Table 2 lists several exemplary 3V class redox shuttles reported in the open literature [18-23].

The power of redox shuttles lies on the fact that their molecules continuously move back and forth ("shuttle") between the positive and negative electrodes to carry a large amount of charge through the cell in a short period of time; this mechanism was fully demonstrated by Bard et al. using a scanning electrochemical microscope (SECM) to achieve single molecule detection [24]. A small electrolyte droplet containing roughly 1 molecule of [(trimethylammonio)methyl]ferrocene was sandwiched between an indium-tin oxide (ITO) glass and a SECM tip with a gap of 10 nm. It has been demonstrated that the single molecule of [(trimethylammonio) methyl]ferrocene can carry about 1 pA current from the SECM tip to the ITO counter electrode with an amplification factor of about 10^7. This means that the redox shuttle molecule has to complete 10^7 oxidation-reduction reaction cycles every second. If the investigated molecule has a probability of 1 part per million to undergo irreversible decomposition for each redox cycle, then a simple calculation shows that the possibility of the redox shuttle to survive after 1 second of testing is about 45 part per million. Therefore, an ideal redox shuttle for lithium-ion batteries should be extremely stable to offer hours, or even hundreds of hours, of overcharge protection. In addition, an ideal redox shuttle is also expected to have a redox potential about 0.3–0.4 V higher than the normal operation potential of the positive electrode to minimize the leakage current, or self-discharge current, described as Equation 3.

$$i = i_0 \exp\left(\frac{\varepsilon - E^o}{RT}\right) \tag{3}$$

In Equation 3,

i = self discharge current density,

i_0 = exchange current density that is a characteristic kinetic constant for a given electroactive specie,

ε = the potential of the working electrode, and

E^o = the standard redox potential of the electroactive specie.

Finally, a good solubility and high diffusion coefficient of the redox shuttles in non-aqueous electrolytes are also highly desired to maximize their mobility through the cell.

Stability of Redox Shuttles

Electronic Stability

After the first commercialization of lithium-ion batteries using $LiCoO_2$ as the positive electrode material, Adachi et al. proposed that aromatic compounds can be the 4V redox shuttles for applications in state-of-the-art lithium-ion batteries [25, 26]. They also showed that aromatic compounds with two methoxyl substitution groups could have a better stability than their unsubstituted counterparts [25]. Using 1,4-dimethoxybenzene as an example, the authors believed that the added methoxy-substituted groups at the 1,4 positions helped provide more resonance structures to stabilize the doubly oxidized cation (see Figure 2). This mechanism can also directly apply to 1,2-dimethoxybenzene. However, doubly oxidized 1,3-dimethoxybenzene shows less stable resonance structures and hence the 1,3 substitution configuration is less favored.

Apparently, the resonance structures can explain the relative stability of dimethoxybenzene with different substitution configurations. However, it will be highlighted in the following pages that the redox shuttles for overcharge protection in general only involve the first oxidation reaction, and the doubly oxidized dimethoxybenzene generally undergoes a very fast irreversible decomposition reaction. Therefore, the second oxidation reaction is generally not involved in a working battery. Moreover, the relative stability of a doubly oxidized cation might not be a good indicator for the stability of singly oxidized cations.

It is believed that the need for two methoxyl groups comes from the aromaticity, or the 4n+2 rule, of unsaturated hydrocarbons [27]. When the number of electrons in the flat p-π molecular orbital equal 4n+2 (n is an integer), the molecule possesses aromaticity and the charged structures can be stabilized by delocalizing the charge in the big π orbital, reducing the net charge on a single atom. Dimethoxybenzene is a good example for the 4n+2 rule with 6 π electrons in the benzene ring and one pair of p-electrons from each conjugated O atom.

Figure 3 shows the molecular configurations of 1,4-dimethoxybenzene and its corresponding radical cation, which is generated after 1,4-dimethoxybenzene is oxidized (losing one electron). Both configurations are presented according to the full geometry optimization using ab initio calculations. The most significant difference between these configurations is the tetrahedral angle between the methoxyl groups and the aromatic plane. At its reduced state, 1,4-dimethoxybenzene has its methoxyl groups pointing out of the aromatic plane (see Figure 3a). However, the methoxyl groups stay in the aromatic plane when the molecule is oxidized to form a radical cation (see Figure 3b). Figure 3b also shows a slight increase of the bond order between the carbon atom (on the aromatic ring) and the oxygen atom.

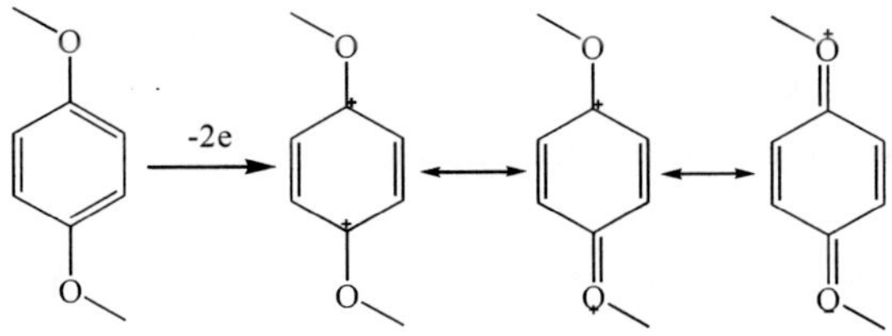

Figure 2. Canonical structures of 1,4-dimethoxybenzene after two-electron oxidation.

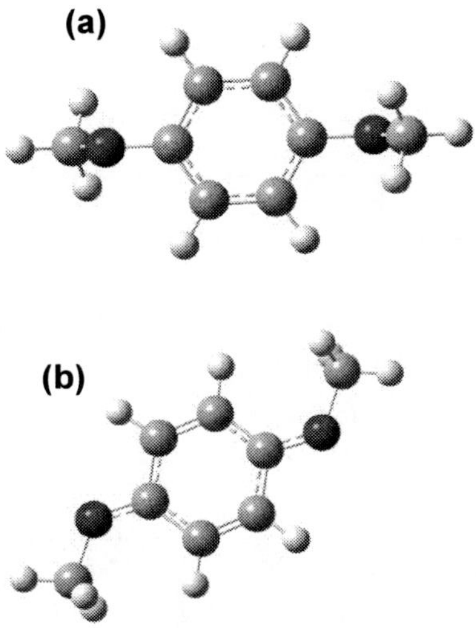

Figure 3. Optimized geometry of 1,4-dimethoxybenzene in its (a) reduced state and (b) oxidized state.

The above evidence indicates some sort of interaction between the aromatic ring and the oxygen atom, and the significance of the interaction needs to be verified. Thus, the geometry of the transition state of both the reduced state and the radical cation was also optimized by fixing one of the methoxyl groups in or out of the aromatic plane. The transition state mentioned above is for the rotation of the methoxyl group with the C-O bond. The energy of each state was determined by density function theory. The energy differences between the in-plane configuration and the out-of-plane configuration are listed in Table 3. Apparently, the energy difference is the energy barrier needed to rotate the C(ph)-O σ bond from an out-of-plane configuration to an in-plane configuration. The energy barrier for the reduced state is about -11.2 kJ/mol, which is acceptable for the rotation of the σ bond. When the molecule is oxidized, the energy barrier jumps to about 52.7 kJ/mol, which is unusually high for the simple rotation of the σ bond.

Table 3. Energy barriers for C-X bond rotation.

Chemical Name	Chemical Structure	Energy Barrier for Reduced State (kJ/mol)	Energy Barrier for Oxidized State (kJ/mol)
1,4-dimethoxybenzene (X=O)	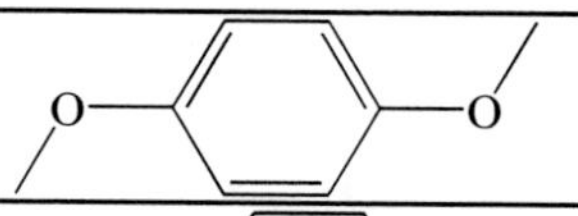	-11.2	52.7
1,4-dimethyl benzene (X=C)	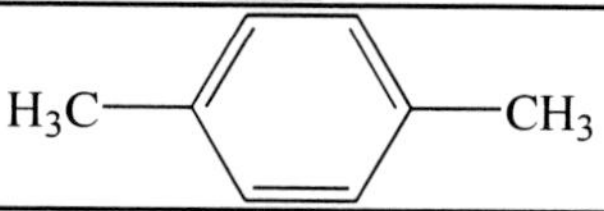	0.2	3.1
1,4-dimethylthio benzene (X=S)		2.3	71.2

(Reproduced from *J. Electrochem. Soc.*, 153: A2215, Copyright (2006), with permission from the Electrochemical Society.)

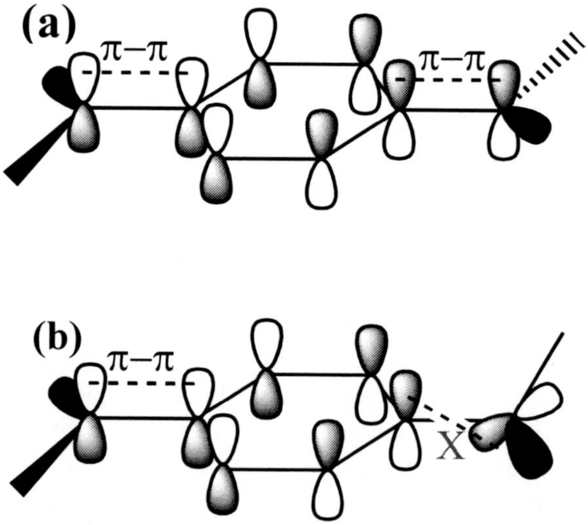

(Reprinted from *J. Electrochem. Soc.*, 153: A2215, Copyright (2006), with permission from the Electrochemical Society.)

Figure 4. Schematic illustration of π-π interaction between the aromatic ring and the substitution groups.

Figure 4a schematically shows the highest occupied molecular orbital (HOMO) configuration of the oxidized 1,4-dimethoxybenzene. The oxygen atom in Figure 4a adopts an sp^2 hybrid configuration with one O-C σ bond, one full sp^2 orbital staying in the aromatic plane, and one full p-orbital rotating perpendicular to the aromatic plane. Obviously, the p-orbital of the oxygen atom cannot receive more electrons, but it can partially donate its electrons to the aromatic ring through the π-π interaction, as shown in Figure 4a. When the molecule is not oxidized, there is no formal charge bearing on the aromatic ring, so that the electron donation seems insignificant and the π-π interaction is weak. Thus, the C(ph)-O bond has more σ characteristics and has a small energy barrier for rotation. When the molecule is oxidized, the donation of an electron from the oxygen atom to the aromatic ring becomes significant because the aromatic ring bears a +1 formal charge. If one of the methoxyl groups rotates by 90°, as shown in Figure 4b, the p-orbital of the oxygen atom will stay in the aromatic plane and the π-π interaction will be deconstructed, which is believed to contribute to the high rotation energy barrier of the radical cation.

Table 3 also lists the rotation energy barriers of 1,4-dimethyl benzene, which lacks π-π interaction between the aromatic ring and the substitution groups because of the sp^3 configuration of carbon atoms in the substitution groups. As expected, the rotation energy barrier for both reduced and oxidized 1,4-dimethybenzene is very small because of the lack of π electrons in the substitution groups. Another typical example is 1,4-dimethythio benzene, which has very similar electronic structure to 1,4-dimethoxybenzene. The 3p electrons on S can interact with the aromatic ring as the 2p electrons on O do. Therefore, a high rotation energy barrier is also predicted by the theoretical calculation for the oxidized state of 1,4-dimethythio benzene, as shown in Table 3.

Given the significance of the lone p electrons on the substitution groups, there is a question about the stability of V-group-based (such as N) substituted aromatic compounds. In order to answer this question, we tested some commercially available aromatic compounds comprising N in the full cell configuration. Such compounds proved to be very unstable as redox shuttles. A reasonable explanation is that the amine group, a Lewis base, is susceptible to attacking by the Lewis acid like Li^+. Once the amine group is attacked, the N atom is forced to adopt a sp^3 configuration and lose the π-π interaction with the aromatic ring. (Note: $-OCH_3$ can also be attacked by H^+ or Li^+.) Group VI elements, such as O or S, have two lone pairs of electrons and can maintain the

sp^2 configuration even after being attacked by the Lewis acid (i.e., H^+ or Li^+), presuming that the possibility of accepting more than one Lewis acid can be ignored. Therefore, some S-based aromatic compounds were also reported as redox shuttles for overcharge protection [28-30].

Structural Stability

Although stabilized by two methoxyl groups, 1,4-dimethoxybenzene has been reported undergoing a radical polymerization reaction when being oxidized [31], as shown in Equation 4.

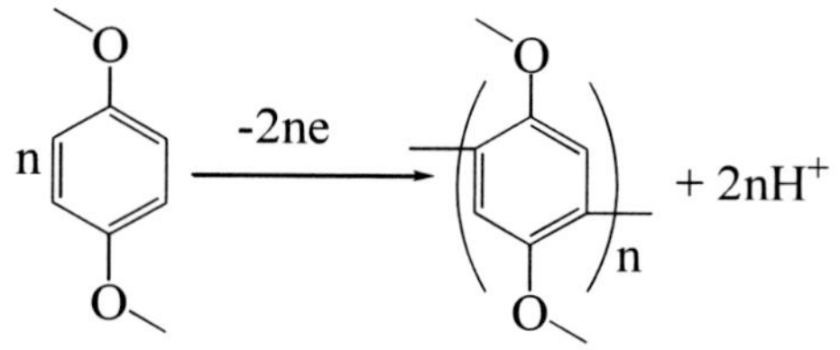

(4)

It is speculated that the polymerization mechanism in Eq. 4 involves two steps. Once 1,4-dimethoxybenzene is oxidized, a radical cation will be formed, and the radical cation can be further stabilized by losing a proton with no charge bearing on the aromatic ring (see Equation 5). The resulted radical can attack other 1,4-dimethoxybenzene molecule as shown in Equation 6. After the polymerization reaction, 1,4-dimethoxybenzene will be converted into a polymer and lose shuttle capability. Hence, 1,4-dimethoxybenzene cannot be a reversible redox shuttle for lithium-ion batteries.

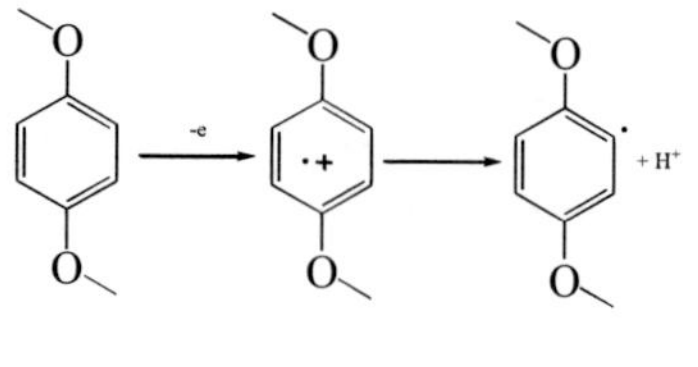

(5)

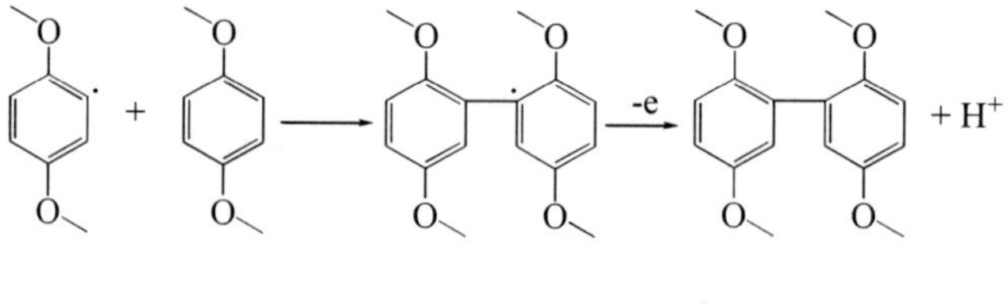

(6)

Adachi et al. proposed that halogenated dimethoxybenzene could be a stable redox shuttle for overcharge protection of 4V class lithium-ion batteries [25, 26], with 1-bromo-2,5-dimethoxybenzene identified as most stable [25]. However, Dahn et al. reported that 1-bromo-2,5-dimethoxybenzen is not stable at all [14]. Figure 5 shows the voltage profile of a $LiFePO_4$/graphite lithium-ion cell during the overcharge test. The electrolyte used was 0.08 M 1-bromo-2,5-dimethoxybenzene and 0.7 M lithium bis(oxalato)borate (LiBOB) in a mixture solvent of propylene carbonate (PC) and dimethyl carbonate (DMC) with a volume ratio of 1:2. When the cell was initially charged, its voltage increased normally with the charge time. The cell voltage surged up sharply after the cell was fully charged at about 3.7 V until the cell voltage reached about 4.3 V, at which point the redox shuttle mechanism of 1-bromo-2,5-dimethoxybenzene was activated. Figure 5 clearly shows that 1-bromo-2,5-dimethoxybenzene can provide a certain amount of overcharge protection, but the overcharge protection disappeared after 6 cycles with 100% overcharge for each cycle [14].

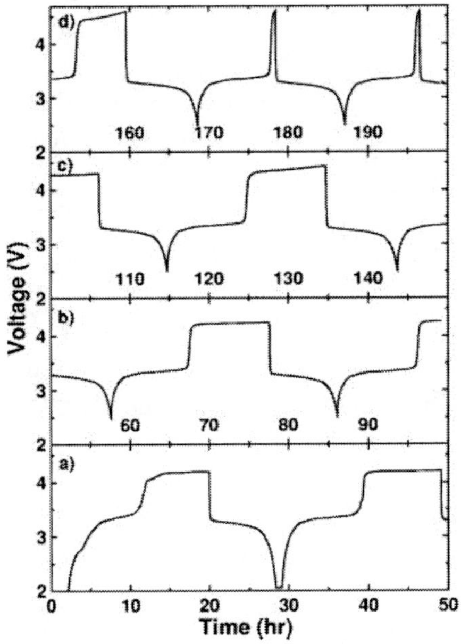

(Reprinted from *Electrochem. Solid-State Lett.*, 8(1):159, Copyright (2005), with permission from the Electrochemical Society.)

Figure 5. Cell potential *vs.* time for a $LiFePO_4$/graphite cell containing a 1-bromo 2,5-dimethoxybenzene shuttle additive. The electrolyte salt was 0.7 M LiBOB. The x axis for (a) covers the time periods of 0–50 h, (b) 50–100 h, (c) 100–150 h, and (d) 150–200 h.

As a comparison, Figure 6 shows the voltage profile of another $LiFePO_4$/graphite lithium-ion cell using 0.08 M 1,4-ditertbutyl-2,5-dimethoxybenzene as the redox shuttle instead of 1-bromo-2,5-dimethoxybenzene. Under the same testing condition, the redox shuttle mechanism of the cell with 1,4-ditertbutyl-2,5-dimethoxybenzene was still fully functional after more than 120 cycles [14]. The instability of 1-bromo-2,5-dimethoxybenzene as a redox shuttle can be simply attributed to (1) 1,4-dimethoxybenzene itself is not stable after oxidation, as shown in Equation 4 and (2) substitution of electron withdrawing group, such as a halogen, on the benzene ring will increase the positive atomic charge bearing on the aromatic ring and ease the initiation of the polymerization reaction (Equation 5). The density function theory calculation shows that the energy needed for reaction (5), cleavage of C-H bond in the singly oxidized 1,4-dimethoxybenzene, is about 1,200 kJ/mol. The energy needed for 1,4-difluoro-2,5-dimethoxybenzene is about 1,000 kJ/mol while the energy needed for 1,4-ditertbutyl-2,5-dimethoxybenzene is about 1,250 kJ/mol [27]. Therefore, it can be concluded that the partial substitution of hydrogen in the aromatic ring can have only a limited impact on the deprotonation reaction (Eq. 5). Therefore, the exemplary stability of 1,4-ditertbutyl-2,5-dimethoxybenzene cannot simply be explained by the stabilization effect of the tertbutyl groups that donate an electron. A reasonable speculation is that the bulky terbutyl groups prevent the 1,4-ditertbutyl-2,5-dimethoxybenzene and its radical cation from attacking each other and hence suppress the polymerization reaction as shown in Equation 6 [14].

In order to accelerate the search for stable redox shuttles, Dahn et al. established a computational method to roughly predict the stability of redox shuttles [32]. In Dahn's method, the binding energy (Eb) of an ethyl radical and candidate redox shuttle molecule is calculated; a low binding energy is expected to correlate to a higher stability. The computational results of this method agreed well with the previous experimental results reported by Dahn's group; 2,5-tert-butyl-1,4-dimethoxybenzene [14, 33, 34] is the most stable one, followed by 10-methylphenothiazine [28, 33] and 2,2,6,6,-tetramethylpiperdine-1-oxyl [33, 35]. By examining Equation 6, one can quickly find that Dahn's method basically uses the ethyl radical to mimic the radical generated after a redox shuttle molecule is oxidized (see Equation 5) to reduce the computational cost. However, there are still several decomposition reactions that cannot be predicted with this mechanism. For example, fully substituted aromatic compounds [27], such as 2,3,5,6-tetrafluoro-1,4-ditertbutybenzene [36], are unlikely to generate a free radical. There are also circumstances in which the decomposition reaction is an SN1 reaction, such as the cleavage of C-O bond in the alkoxy groups [36].

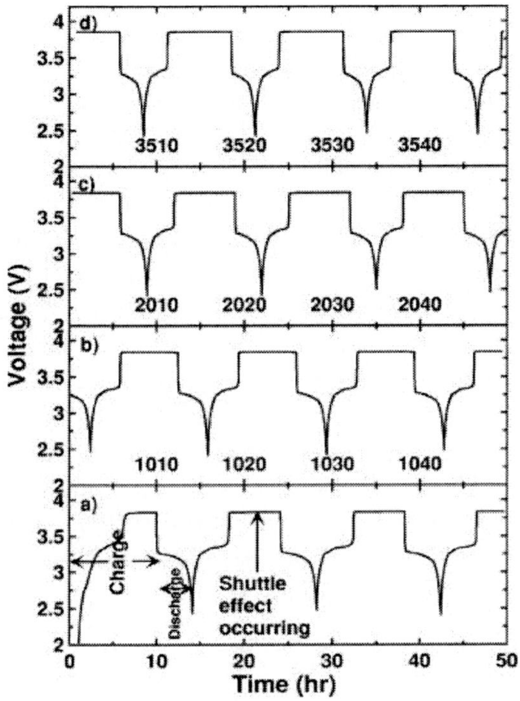

(Reprinted from *Electrochem. Solid-State Lett.*, 8(1):159, Copyright (2005), with permission from the Electrochemical Society.)

Figure 6. Cell potential vs. time for a $LiFePO_4$/graphite cell containing a 2,5-ditertbutyl-1,4-dimethoxybenzene shuttle additive. The electrolyte salt was 0.7 M LiBOB. The x axis for (a) covers the time period of 0–50 h, (b) 1,000–1,050 h, (c) 2,000–2,050 h, (d) 3,500–3,550 hours.

Figure 7 shows the results of ab initio calculations on the decomposition pathway of 2,3,5,6-tetrafluoro-1,4-ditertbutybenzene after oxidation. The ab initio calculation shows that 2,3,5,6-tetrafluoro-1,4-ditertbutybenezene can be oxidized at a potential of about 4.32 V vs. Li^+/Li, which agrees well with the experimental results as shown in Figure 8a. The ab initio calculation also shows that 2,3,5,6-tetrafluoro-1,4-ditertbutybenzene radical cation (singly oxidized) will spontaneously lose a tertbutyl radical and form a semiquinone. The semiquinone will then be further oxidized at 3.43 V vs. Li^+/Li, which is much lower than the first oxidation potential. The ab initio calculation clearly shows that the final product is 2,3,5,6-tetrafluoroquinone after losing a second tertbutyl group. These theoretical results were also confirmed with experimental data, as shown in Figure 8.

Figure 8a shows that one irreversible oxidation peak was observed with the onset potential at about 4.1 V vs. Li^+/Li. Once the irreversible oxidation reaction occurred, an irreversible reduction peak was observed at about 3.0 V vs. Li^+/Li accordingly. After the electrolyte was tested up to 4.6 V vs. Li^+/Li for several cycles, a small reduction peak was observed at about 2.96 V vs. Li^+/Li even when the upper cutoff potential was set to 4.0 V vs. Li^+/Li (see black line in Figure 8b). In order to confirm the degradation pathway, a small amount, but not measured, of 2,3,5,6-tetrafluoroquinone (TFQ) was added to the test electrolyte, and a significant intensity increase of the peak at 2.96 V was observed (see Figure 8b). This finally confirmed that the product of the decomposition reaction is TFQ [36]. Another reported example that undergoes similar decomposition is 2,5-di-tertbutyl-1,4-dimethoxybenzene. Although 2,5-di-tertbutyl-1,4-dimethoxybenzene has been reported as a stable redox shuttle [13, 14, 33, 34, 37, 38], it can also undergo an irreversible decomposition reaction after being doubly oxidized [36, 39]. It is believed that 2,5-di-tertbutyl-1,4-dimethoxybenezene is stable up to singly oxidized state, but can lose one or two methyl groups to form semiquinone or quinone after being doubly oxidized.

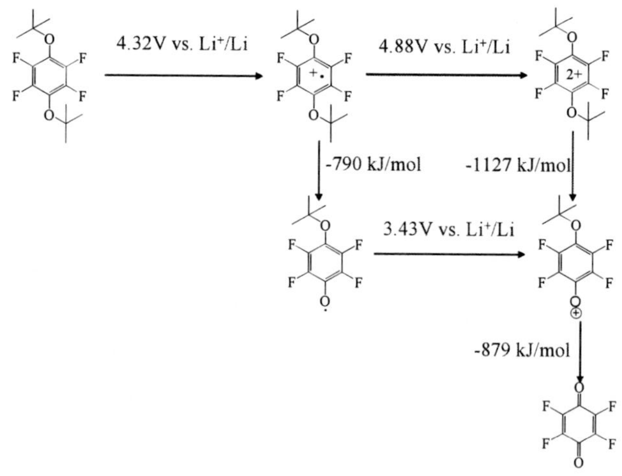

Figure 7. Schematic of decomposition pathway of oxidized 2,3,5,6-tetrafluoro-1,4-di-tertbutybenzene.

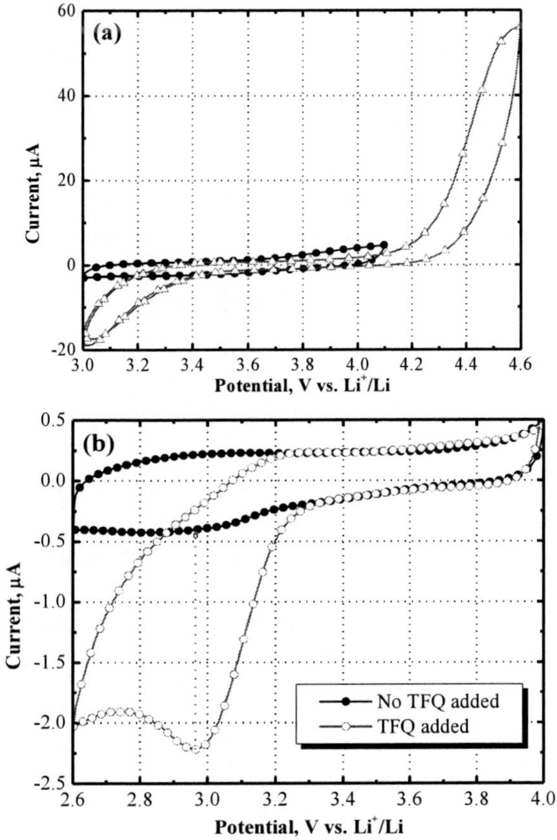

(Reprinted from *Electrochim. Acta,* Vol 53/2: 453-458, Copyright (2007), with permission from Elsevier)

Figure 8. (a) Cyclic voltammograms of 50 mM 2,3,5,6-tetrafluoro-1,4-di-tert-butoxybenzene and 1.0 M $LiPF_6$ in EC/EMC (3:7, by weight), and (b) Cyclic voltammograms of 50 mM 2,3,5,6-tetrafluoro-1,4-di-tert-bytoxybenzene without and with the addition of 2,3,5,6-tetrafluorquinone (TFQ).

Examples of Stable Redox Shuttles

Aromatic Redox Shuttles

The discussion so far has demonstrated that redox shuttles can have multiple decomposition pathways, making the design of a stable redox shuttle very difficult. Several compounds have been reported in the literature as stable redox shuttles for overcharge protection of lithium-ion batteries [13-17, 28, 35, 40-42], and they are useful models for establishing guidelines for the search for the next stable redox shuttle.

Aromatic compounds were proposed as promising redox shuttles for overcharge protection of 4V class lithium-ion batteries about a decade ago. However, the first stable aromatic redox shuttle was recently reported by Dahn et al. [14], in which work 2,5-di-tertbuty-1,4-dimethoxy, whose redox potential is about 3.96 V vs. Li^+/Li, showed unprecedented stability as a redox shuttle for lithium-ion cells (see Figure 9). Figure 9 shows the charge and discharge capacity of a $LiFePO_4$/graphite lithium-ion cell comprising 0.08 M 2,5-di-tertbuty-1,4-dimethoxybenzene during overcharge test, and the voltage profile of this cell is shown in Figure 6. It is shown that the cell was charged for about 290 mAh/g and delivered about 145 mAh/g during discharge. The difference between the charge capacity and the discharge capacity was caused primarily by the overcharge of the cell (~100% overcharge) and was handled by the added redox shuttle, 2,5-di-tertbutyl-1,4-dimethoxybenzene. The overcharge protection mechanism survived for more than 260 cycles—in other words, the cell had tolerated more than 260 times of its practical capacity before the overcharge protection mechanism failed.

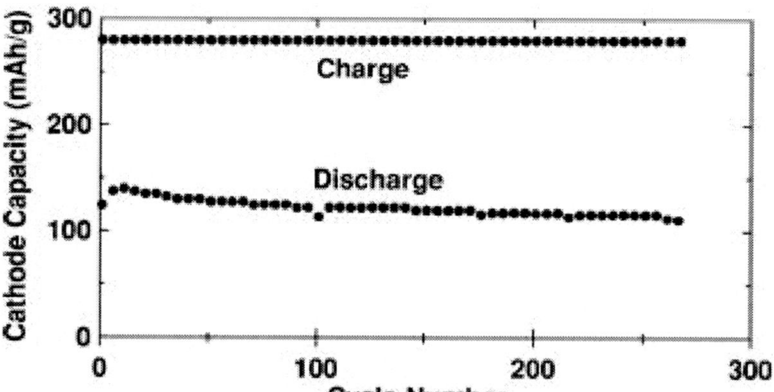

(Reprinted from *Electrochem. Solid-State Lett.*, 8(1):159 (2005), Copyright (2005), with permission from the Electrochemical Society.)

Figure 9. Positive electrode specific capacity *vs.* cycle number for a $LiFePO_4$/graphite cell charged and discharged at a C/2 rate at 30°C. A 0.7 M LiBOB-based electrolyte containing 0.08 M 2,5-ditertbutyl-1,4-dimethoxybenzene was the shuttle. Each recharge was for a constant capacity selected to be 200% of the nominal cell capacity. Each discharge was to a 2.5 V cutoff potential.

Another aromatic redox shuttle reported by Dahn et. al. is 10-methylphenothiazine (MPT), which has a redox potential of about 3.47 V vs. Li^+/Li. Figure 10 shows (a) the voltage profile and (b) specific charge/discharge capacity of a $LiFePO_4/Li_4Ti_5O_{12}$ lithium-ion cell containing 0.1 M MPT in 0.5 M $LiPF_6$ in a mixture solvent of propylene (PC), dimethyl carbonate (DMC), ethylene carbonate (EC), and diethyl carbonate (DEC) with a volume ratio of 1:2:1:2 as the electrolyte. Under the specific overcharge protocol, about 160% overcharge for each cycle, the overcharge protection provided by the added MPT survived for more than 200 cycles. Compared to 2,5-di-tertbuty-1,4-dimethoxybenzene, MPT would be expected to be very unstable because of (1) the stabilization effect of N-based substitution is expected to be much weaker than VI group, such as O- or S-based substitution groups, and (2) the unprotected C-H bond on the aromatic ring is believed to be susceptible to attack by radicals (see equations 5 and 6). The surprising stability of MPT (see Figure 10) can be explained by its low redox potential, which is about 0.5 V lower than that of 2,5-ditert-butyl-1,4-dimethoxybenzene. A speculation on the stability of MPT is that MPT absorbed less energy to be oxidized into a radical cation, and hence the possibility of C-H cleavage to form a radical (Equation 5) is lower than that of 2,5-di-tertbuty-1,4-dimethoxybenzene.

The redox shuttles mentioned above have a redox potential lower than 4.0 V vs. Li^+/Li, and are only practical for overcharge protection of lithium-ion cells using $LiFePO_4$ as the positive electrode. However, the materials commonly used in current lithium-ion batteries generally have a higher working potential (>4.0 V vs. Li^+/Li). These materials are primarily $LiCoO_2$ and a small portion of $LiMn_2O_4$ and $Li[Mn_{1-x-y}Ni_xCo_y]O_2$. Therefore, there is a need to develop redox shuttles with a redox potential of 4.2 V vs. Li^+/Li or higher.

Dahn et. al. have shown that the redox potential of fluorinated 2,2,6,6-tetramethylpiperinyl-oxides (TEMPO) increases with the degree of fluorination [35]. It is well known that the substitution of an electron-withdrawing group such as F around the redox active site, -N-O radical in the case of TEMPO, can reduce the electron density on the active site and increase the energy needed to withdraw one electron out of the active site (oxidation process). Therefore, substitution of electron withdrawing groups can be an effective way to increase the redox potential of aromatic redox shuttles for applications in highe-voltage systems. It is also known that the substitution of electron-withdrawing groups will increase the positive charge density at the active site after it is oxidized. In the case of fluorinated dimethoxybenzene, the substitution of strong electron-withdrawing groups like F will facilitate the deprotonation reaction (Equation 5) and hence enhance the decomposition or polymerization of aromatic redox shuttles [27]. Therefore, full substitution is a natural strategy to take advantage of both higher redox potential with electron-withdrawing substitution and the elimination of the post-oxidation deprotonation reaction [27]. The discovery of 2-(pentafluorophenyl)-tetrafluoro-1,3,2-benzodioxaborole (PFPTFBB) as a stable redox shuttle is an example to apply the full substitution strategy [17].

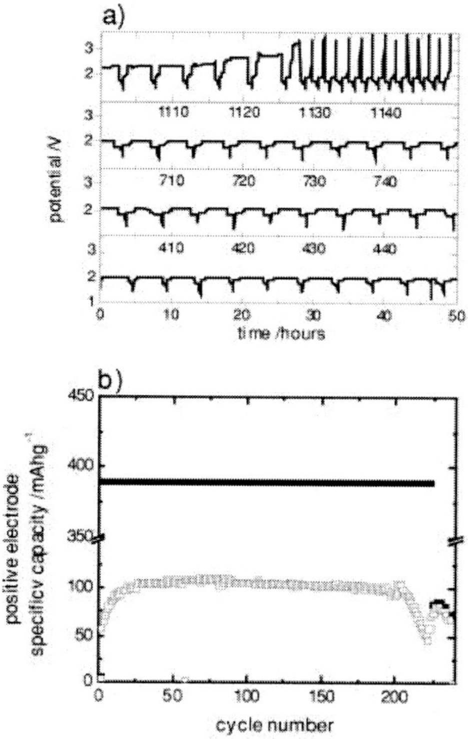

(Reprinted from *J. Electrochem. Soc.*, 153: A288, Copyright (2006), with permission from the Electrochemical Society.)

Figure 10. (a) Potential vs. time for a $LiFePO_4/Li_4Ti_5O_{12}$ coin cell containing 0.1 M MPT in 0.5 M $LiPF_6$ electrolyte solution charged and discharged at a nominal C/2 rate. Selected cycles are shown. (b) Positive electrode specific capacity vs. cycle number for the same cell.

Figure 11 shows cyclic voltammograms of 0.05 M PFPTFBB and 1.2 M $LiPF_6$ in ethylene carbonate (EC)/ethyl methyl carbonate (EMC) (3:7, by weight). Figure 11 shows that PFPTFBB has a reversible redox reaction at about 4.43 V vs. Li^+/Li. The onset potential of PFPTFBB is about 4.3 V vs. Li^+/Li, which is high enough to provide overcharge protection for most state-of-the-art positive electrode materials. Figure 12 shows the charge/discharge capacity of a graphite/$LiNi_{0.8}Co_{0.15}Al_{0.05}O_2$ lithium-ion cell containing 5 wt% PFPTFBB in the electrolyte, which was 1.2 M $LiPF_6$ in EC/PC/DMC (1:1:3, by weight). The cell was charged and discharged at a constant current of C/10 (0.2 mA). During the charging process, the cell was charged to 4.95 V or until 4.0 mAh charge was delivered (100% overcharge). The cell was initially tested at 25°C for 20 cycles and was then heated in an oven to 55°C for another 50 cycles to check the stability of the redox shuttle under a more aggressive testing condition. After that, the cell was tested with a higher constant current of C/5 (as shown above the curve in Figure 12). After 50 cycles at 55°C, the overcharge protection of the redox shuttle was maintained and the cell capacity remained very stable, even though the cell was 100% overcharged every cycle. Afterward, the cell was further tested at 25°C and 55°C with a constant current of C/5 and 100% overcharge. The overcharge protection provided by PFPTFBB finally failed after 170 cycles of 100% overcharge. Figure 12 also shows that the cell completely lost its capacity after 125 cycles. However, the redox shuttle mechanism remained active for another 50 cycles at 55°C after the cell failed.

The uniqueness of PFPTFBB is that it also has a boron-based electron-deficient center, which can serve as an anion receptor with the potential to improve the performance of lithium-ion cells [43]. Figure 13 shows the discharge capacity of MCMB/$Li_{1.1}[Ni_{1/3}Co_{1/3}Mn_{1/3}]_{0.9}O_2$ lithium-ion cells with and without the addition of PFPTFBB as the electrolyte additive. The "AR" labeled on the graph stands for anion receptor and specifically for PFPTFBB. It was confirmed that the addition of PFPTFBB to the electrolyte can significantly improve the capacity retention of lithium-ion cells like other anion receptors [43]. The improvement on the cycling performance was believed to be due to (1) the dissolution of LiF, an insulator to both electron and lithium-ion, from the surface of electrode materials by the added anion receptor [43] and (2) enhanced formation of a solid electrolyte interphase on the negative electrode for better protection [43]. Figure 11 shows the CV of PFPTFBB combining with equivalent amount of LiF. Clearly, the redox reaction at about 4.4 V vs. Li^+/Li was maintained after PFPTFBB reacted with

LiF. This confirms that PFPTFBB is a bifunctional electrolyte additive, serving as both redox shuttle and anion receptor [17].

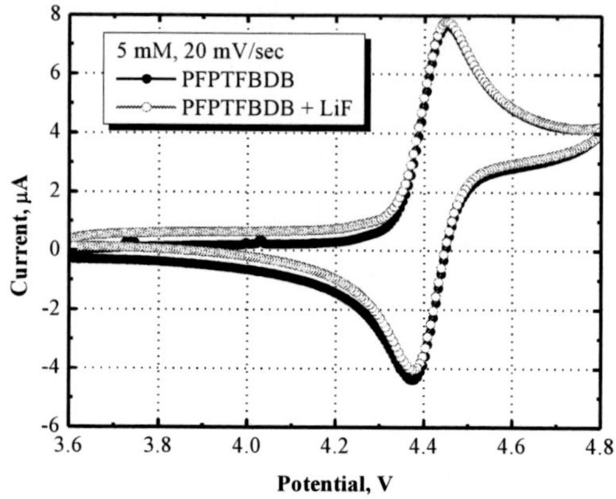

(Reprinted from *Electrochem. Commun.*, Vol 9: 703-707, Copyright (2007), with permission from Elsevier)

Figure 11. Cyclic voltammograms of 0.05 M PFPTFBB and 1.2 M $LiPF_6$ in EC/EMC (3:7, by weight) showing the electrochemical reactivity of PFPTFBB-F^- using a Pt/Li/Li three-electrode cell.

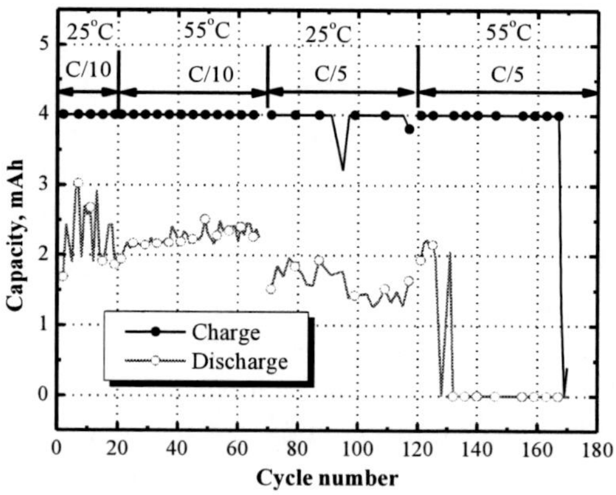

(Reprinted from *Electrochem. Commun.*, Vol 9: 703-707, Copyright (2007), with permission from Elsevier)

Figure 12. Charge and discharge capacity of a graphite/$LiNi_{0.8}Co_{0.15}Al_{0.05}O_2$ lithium-ion cell during the whole course of overcharge test. The electrolyte used contained 5 wt% PFPTFBB.

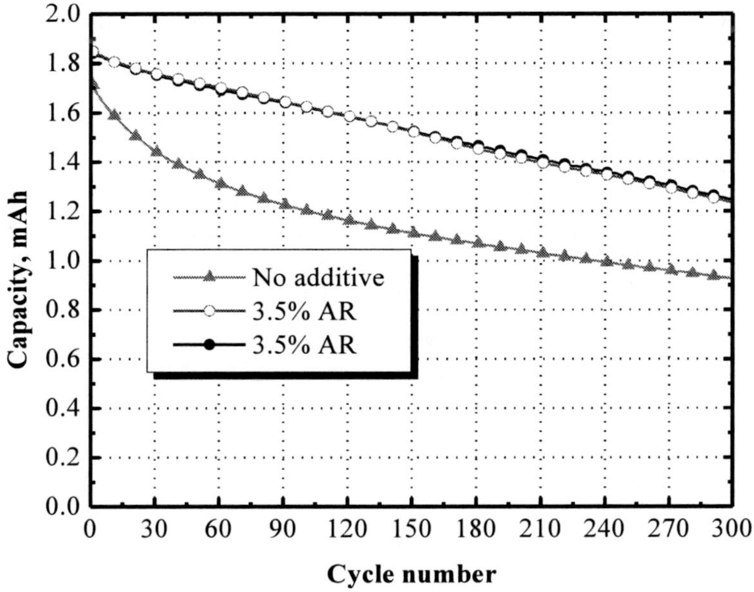

(Reprinted from *Electrochem. Commun.*, Vol 9: 703-707, Copyright (2007), with permission from Elsevier)

Figure 13. Capacity retention of MCMB/$Li_{1.1}[Ni_{1/3}Co_{1/3}Mn_{1/3}]_{0.9}O_2$ cells showing the positive impact of PFPTFBB as an anion receptor. The cells were cycled between 3.0 V and 4.0V with a constant current of C/2 at 55°C.

Non-Aromatic Redox Shuttles

In addition to the aromatic redox shuttles discussed above, several non-aromatic compounds have been reported capable of providing long-term overcharge protection for lithium-ion cells [15, 35, 41, 44-48]. TEMPO is an unusually stable neutral radical that has been extensively investigated in medicinal and biological research. Recently, Dahn et al. reported that TEMPO is capable of reversible redox reaction at 3.52 V vs. Li^+/Li [35, 41], and suggested TEMPO as redox shuttle for overcharge protection of $LiFePO_4$-based lithium-ion cells. Figure 14 shows the capacity vs. cycle number of a $Li_4Ti_5O_{12}$/$LiFePO_4$ lithium-ion cell that was cycled for 120 cycles. The electrolyte tested was 0.3 M TEMPO and 0.5 M LiBOB in PC/DMC/EC/DEC (1:2:1:2 by volume), and the cell was tested with a constant current of C/5. During the whole course of overcharge testing, the excess charge shuttled by

TEMPO was about 20 Ah, which is about 160 times the practical capacity of the cell.

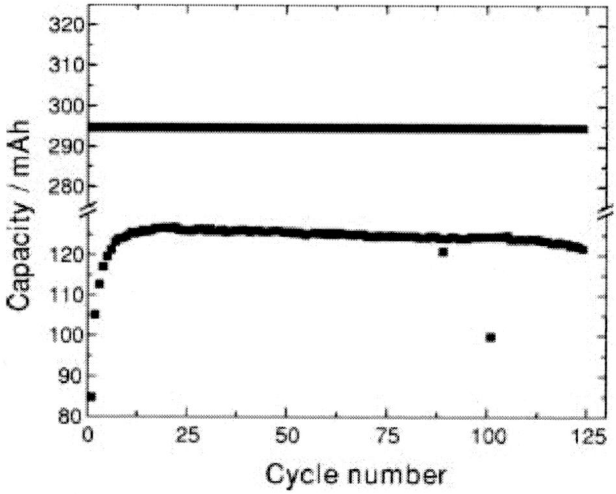

(Reprinted from *J. Electrochem. Soc.*, 153: A1800, Copyright (2006), with permission from the Electrochemical Society.)

Figure 14. Charge and discharge capacity of a $Li_4Ti_5O_{12}/LiFePO_4$ coin cell with 0.3 M TEMPO in a 0.5 M LiBOB electrolyte solution charged and discharged with a current of C/5. Each charge was 20 h long, leading to about 15 h of shuttle-protected overcharge during each cycle.

Another class of non-aromatic redox shuttles is lithium borate cluster salts, $Li_2B_{12}F_{12-x}H_x$(X=1,2,...12) [15, 44], whose redox potential ranges from 4.2 V to 4.7 V vs. Li^+/Li depending on the degree of fluorination. For instance, $Li_2B_{12}F_{12}$ has a redox potential of 4.74 V vs. Li^+/Li and is capable of providing overcharge protection for all state-of-the-art positive electrode materials. The biggest advantage of this class of redox shuttles is that they are actually lithium salts and have much higher solubility in the non-aqueous solvent than the organic redox shuttles. It was observed that lithium borate cluster salts are extreme stable; thermal decomposition of the salts was not observed below 400°C, and they also showed great compatibility with moisture. These unique features are very promising for batteries using HF-sensitive materials like $LiMn_2O_4$ [49].

Most importantly, $Li_2B_{12}F_{12-x}H_x$ has an intrinsic redox potential above 4.2 V vs. Li^+/Li with great reversibility, and hence it is also very promising to serve as a redox shuttle for overcharge protection of 4V class lithium-ion batteries. Figure 15 shows the charge/discharge capacity of a $Li_{1.1}[Mn_{1/3}Ni_{1/3}Co_{1/3}]_{0.9}O_2$/graphite lithium-ion cell using a $Li_2B_{12}F_9H_3$ based non-aqueous electrolyte. The cell had a nominal capacity of 1.6 mAh, and was charged for 3.2 mAh for each cycle to apply about 100% overcharge to the cell. The current used for overcharge testing was 0.53 mA (C/3). Figure 15 shows that the overcharge protection mechanism was functional for about 90 cycles.

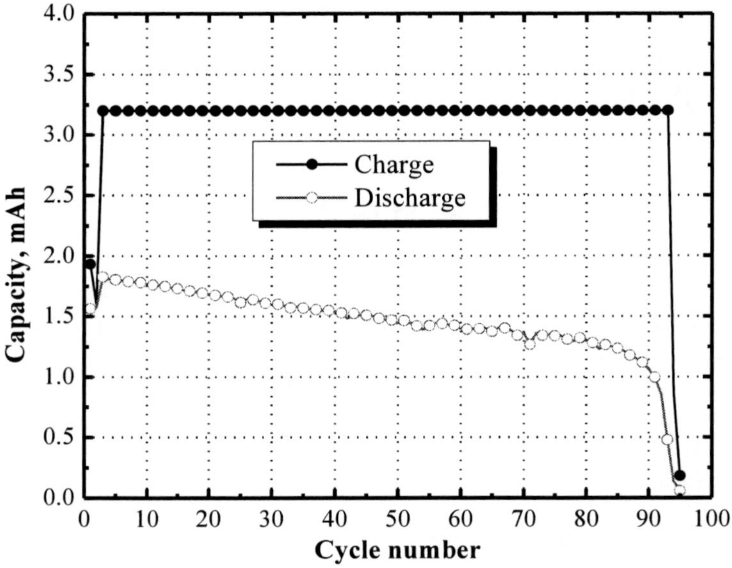

Figure 15. Charge/discharge capacity of a $Li_{1.1}[Mn_{1/3}Ni_{1/3}Co_{1/3}]_{0.9}O_2$/graphite lithium-ion cell using a $Li_2B_{12}F_9H_3$ based non-aqueous electrolyte. The cell had a nominal capacity of 1.6 mAh, and was charged for 3.2 mAh for each cycle to apply about 100% overcharge on the cell. The current used for overcharge test was 0.5 mA (C/3).

CONCLUSION

Overcharging a lithium-ion cell is a severe abuse that can lead to the thermal runaway and catastrophic failure of a battery pack. Redox shuttles are intrinsic chemical overcharge protection mechanism for lithium-ion cells. Several stable redox shuttles have been identified to provide long-term overcharge protection, without the drawbacks of other technologies, which include complex electronic control systems for external voltage regulation and permanent cell loss for deactivation agents. Combining theoretical and experimental investigation on the degradation pathways of redox shuttles can be the key for designing new stable redox shuttles with better performance.

ACKNOWLEDGMENT

Research was supported by U.S. Department of Energy, FreedomCAR and Vehicle Technologies Office. Argonne National Laboratory is operated for the U.S. Department of Energy by UChicago Argonne, LLC, under contract DE-AC02-06CH11357.

REFERENCES

[1] Oh, B; Amine, K. *Solid State Ionics*, 2004, 175 (1-4), 785-788.
[2] Miyashiro, H; Seki, S; Kobayashi, Y; Ohno, Y; Mita, Y; Usami, A. *Electrochem. Commun.*, 2005, 7 (11), 1083-1086.
[3] Kobayashi, Y; Seki, S; Yamanaka, A; Miyashiro, H; Mita Y; Iwahori, T. *J. Power Sources*, 2005, 146 (1-2), 719-722.
[4] Ohsaki, T; Kishi, T; Kuboki, T; Takami, N; Shimura, N; Sato, Y; Sekino M; Satoh, A. *J. Power Sources*, 2005, 146 (1-2), 97-100.
[5] Leising, RA; Palazzo, MJ; Takeuchi ES; Takeuchi, KJ. *J. Electrochem. Soc.*, 2001, 148 (8), A838-A844.
[6] Leising, RA; Palazzo, MJ; Takeuchi ES; Takeuchi, KJ. *J. Power Sources*, 2001, 97-8 681-683.
[7] Wang, H; Chen, MC. *Electrochem. Solid-State Lett.*, 2006, 9 (2), A82-A85.
[8] Yamanaka, Y; Miki, K. US Pat. No. 6642694
[9] Nakamura, S. US Pat. No. 6697259B1
[10] Mao, H; Sacken, UV. US Pat. No. 6033797
[11] Reimers, JN; Way, BM. US Pat. No. 6074777
[12] Mao, H. US Pat. No . 5879834
[13] Dahn, JR; Chen, J. US Pat. Appl. No. 20050221168
[14] Chen, J; Dahn, JR. *Electrochem. Solid-State Lett.*, 2005, 8 (1), A59-A62.
[15] Chen, Z; Amine, K. US Pat. Appl. No. 20070072085
[16] Amine, K; Chen Z; Wang, Q. US Pat. App. No. 20060199080
[17] Chen, Z; Amine, K. *Electrochem. Commun.*, 2007, 9 703-707.
[18] Behl, WK; Chin, DT. *J. Electrochem. Soc.*, 1988, 135 (1), 21-25.
[19] Behl, WK; Chin, DT. *J. Electrochem. Soc.*, 1988, 135 (1), 16-21.

[20] Behl, WK. *J. Electrochem. Soc.*, 1989, 136 (8), 2305-2310.
[21] Abraham, KM; Pasquariello DM; Willstaedt, EB. *J. Electrochem. Soc.*, 1990, 137 (6), 1856-1857.

[22] Golovin, MN; Wilkinson, DP; Dudley, JT; Holonko D; Woo, S. *J. Electrochem. Soc.*, 1992, 139 (1), 5-10.
[23] Narayanan, SR; Surampudi, S; Attia AI; Bankston, CP. *J. Electrochem. Soc.*, 1991, 138 (8), 2224-2229.
[24] Fan, FRF; Bard, AJ. *Science*, 1995, 267 (5199), 871-874.
[25] Adachi, M; Tanaka K; Sekai, K. *J. Electrochem. Soc.*, 1999, 146 (4), 1256-1261.
[26] Adachi, M. US Pt. No. 576119
[27] Chen, ZH; Wang QZ; Amine, K. *J. Electrochem. Soc.*, 2006, 153 (12), A2215-A2219.
[28] Buhrmester, C; Moshurchak, L; Wang, RCL; Dahn, JR. *Journal of the Electrochemical Society*, 2006, 153 (2), A288-A294.
[29] Lee, DY; Lee, HS; Kim, HS; Sun, HY; Seung, DY. *Korean J. Chem. Eng.*, 2002, 19 (4), 645-652.
[30] Abraham, KM; Roham, JF; Foo, CC; Pasquariello, DM. US Pat. No. 5858573
[31] deMartinez, MC; Marquez, OP; Marquez, J; Hahn, F; Beden, B; Crouigneau, P; Rakotondrainibe, A; Lamy, C. *Syth. Metals*, 1997, 88 (3), 187-196.
[32] Wang, RL; Dahn, JR. *J. Electrochem. Soc.*, 2006, 153 (10), A1922-A1928.
[33] Moshurchak, LM; Buhrmester, C; Wang RL; Dahn, JR. *Electrochim. Acta*, 2007, 52 (11), 3779-3784.
[34] Moshuchak, LM; Bulinski, M; Lamanna, WM; Wang, RL; Dahn, JR. *Electrochem. Commun.*, 2007, 9 (7), 1497-1501.
[35] Buhrmester, C; Moshurchak, LM; Wang, RL; Dahn, JR. *J. Electrochem. Soc.*, 2006, 153 (10), A1800-A1804.
[36] Chen, ZH; Amine, K. *Electrochim. Acta*, 2007, 53 (2), 453-458.
[37] Dahn, JR; Jiang, JW; Moshurchak, LM; Fleischauer, MD; Buhrmester, C; Krause, LJ. *J. Electrochem. Soc.*, 2005, 152 (6), A1283-A1289.
[38] Buhrmester, C; Chen, J; Moshurchak, L; Jiang, JW; Wang, RL; Dahn, JR. *J. Electrochem. Soc.*, 2005, 152 (12), A2390-A2399.
[39] Moshurchak, LM; Buhrmester, C; Dahn, JR. *J. Electrochem. Soc.*, 2005, 152 (6), A1279-A1282.
[40] Dahn, JR; Buhrmester, C; Wang, RL; Lamanna, WM. US Pat. Appl. No. 20060263697
[41] Dahn, JR; Buhrmester, C. US Pat. Appl. No. 20060263695
[42] Amine, K; Chen, Z; Wang, Q. US Pat. Appl. No. 20060199080
[43] Chen, ZH; Amine, K. *J. Electrochem. Soc.*, 2006, 153 (6), A1221-

A1225.
[44] Amine, K; Liu, J; Jambunathan, K; Peterson, BK; Dantsin, G. US Pat. Appl. No. 20050227143
[45] Chen, GY; Richardson, TJ. *Electrochem. Solid-State Lett.*, 2004, 7 (2), A23-A26.
[46] Chen, GY; Richardson, TJ. *Electrochem. Solid-State Lett.*, 2006, 9 (1), A24-A26.
[47] Chen, GY; Thomas-Alyea, KE; Newman, J; Richardson, TJ. *Electrochim. Acta*, 2005, 50 (24), 4666-4673.
[48] Xiao, LF; Ai, XP; Cao, YL; Wang YD; Yang, HX. *Electrochem. Commun.*, 2005, 7 (6), 589-592.
[49] Amine, K; Liu, J; Belharouak, I; Kang, SH; Bloom, I; Vissers, D; Henriksen, G. 12th International Meeting on Lithium Batteries. 2004. Nara, JAPAN.

In: Lithium Batteries: Research, Technology… ISBN: 978-1-60741-722-4
Editors: Greger R. Dahlin, et al.

Chapter 5

THERMAL STABILITY AND ELECTROCHEMICAL PERFORMANCE OF $LiCoO_2$ AND $LiCo_{0.2}Ni_{0.8}O_2$ IN LITHIUM-ION BATTERIES

George Ting-Kuo Fey[1] and T. Prem Kumar[2]

[1] Department of Chemical and Materials Engineering, National Central University Chung-Li, Taiwan, R.O.C.

[2] Electrochemical Power Systems Division, Central Electrochemical Research Institute Karaikudi 630006, Tamil Nadu, India.

ABSTRACT

Parallel to the rising market for lithium-ion power packs, more incidents of severely debilitating and sometimes fatal tragedies, as a result of battery hazards are being reported. Some of the safety risks of lithium-ion batteries are inherent in the fact that they combine highly energetic materials that are in contact with a flammable electrolyte based on organic solvents. Moreover, the potential ranges experienced by these cells are beyond the thermodynamic stability windows of the electrolytes, which can decompose upon contact with the charged active materials. The interface between the cathode and electrolyte is of special concern since partial dissolution of the active material can create further complications. This chapter discusses processes at the positive electrode that can lead to thermal runaway, especially at those based on the most popular cathode materials, $LiCoO_2$ and $LiNi_{0.8}Co_{0.2}O_2$. Measures such as coating cathode particles with inert oxides have been shown to improve cell safety by increasing the onset temperature of electrode-electrolyte reactions and lowering the exothermicity of such reactions. Additionally, coatings also bestow improved cyclability to the cathodes. Reactivity of cathode active materials is also related to electrolyte composition. Electrolyte additives and non-flammable electrolytes are a case in point. Techniques for studying thermal stability such as differential scanning calorimetry and accelerating rate calorimetry are also discussed.

1. Introduction

Several instances of exploding lithium-ion battery packs have been reported in recent years, which have raised pertinent questions on the safety of such devices [1]. Faced with the consequences of consumer dissatisfaction, product recalls and market share loss, manufacturers are placing greater emphasis on the safety aspects of lithium-ion batteries. Constructed with highly energetic active materials in contact with a flammable electrolyte based on organic solvents, the batteries can become unsafe if subjected to conditions for which they were not designed, such as overcharging, disposing in fire, external short circuiting and crushing. Moreover, the potential ranges experienced in common lithium-ion cells are beyond the thermodynamic stability windows of the electrolytes, which decompose upon contact with the charged active materials. The cathode/electrolyte interface is further complicated by partial dissolution of the positive active materials [2–4], especially at the end of charging and at elevated temperatures [5–11]. Safety improvements have been made by using safety vents, positive temperature coefficient elements, shutdown separators/additives, oxidation-tolerant or less flammable electrolytes and redox shuttles.

2. Measurement of Thermal Stability

Many materials are prone to exothermic reactions at elevated temperatures. Therefore, understanding the conditions that trigger such reactions is essential for operating devices that use the materials safely. It is difficult to define the highest safe temperature for a material since it is complicated by a number of factors such as amount, geometry, surface area, particle size and availability of oxidant. Experimental methods for assessing the thermal instability/runaway potential are based primarily on microcalorimetric techniques such as differential thermal analysis (DTA), thermogravimetry (TG), differential scanning calorimetry (DSC) and accelerating rate calorimetry (ARC). However, the most common techniques employed for determining the instability or thermal runaway potential of battery chemicals are DSC and ARC.

2.1. Differential Scanning Calorimetry

Differential thermal analysis and thermogravimetry are widely used for first-level thermal hazard evaluation because of their operational simplicity and approach. The techniques provide quantitative data on sensitivity (exothermic onset temperature) as well as severity (heat of decomposition) [12]. Since they are not suitable for large volumes of samples and devices, have poor reproducibility and use non-adiabatic experimental conditions, these methods cannot be broadly applied in assessing thermal safety. However, DSC as a thermal hazard tool is particularly useful for determining the decomposition mechanisms of reactive chemicals because it measures heat flow as a function of temperature or time. This is done by measuring the difference in the power required to maintain the temperatures of reference and test samples. Typically, a differential scanning calorimeter consists of a furnace, sample holders for the test and reference materials, and thermocouples. The furnace is set on a steady ramp. The thermal responses yield information on onset of heat flow (endothermicity or exothermicity), heat of decomposition, melting/boiling point, decomposition kinetics, etc.. DSC is a quick and relatively cheap method for thermal screening of small samples, so it is a popular technique for early stage product development.

2.2. Accelerating Rate Calorimetry

Adiabatic calorimetry is an important technique for the study of self-propagating and thermally-sensitive reactions. As a thermal hazard evaluation technique, ARC is particularly suitable for large samples, including cells. ARC consists of a container that maintains the test sample under adiabatic conditions with respect to its environment. The inner temperature is continuously monitored and the surrounding temperature is suitably adjusted in order to minimize heat flow between the container and its surroundings. ARC operates on a heat–wait–search principle to identify the initiation and progress of exothermic self-heating processes. First, the experiment is initiated under the 'heat' mode, when the temperature of the sample and container are raised to a set value, followed by a 'wait' mode for a fixed duration. During the subsequent 'search' mode, the rate of increase in the temperature of the sample container is monitored. The 'heat–wait–search' procedure is repeated until the system experiences a self-heating rate above a set threshold. When self-accelerating exothermicity is detected, the controls activate to maintain adiabatic conditions, afterwhich any increase in the temperature of the sample can be attributed totally to the heat generated by the test sample. ARC provides information on the rate of self-heating as a function of temperature, variation of temperature with time, variation of pressure with time/temperature, enthalpy of reactions and kinetic parameters. ARC offers more sensitivity than DSC.

3. Hazard Triggers

3.1. Temperature Coefficient of Cell Voltage

The temperature derivative of the free energy change, ΔG, for a reaction to the associated change in entropy, ΔS, can be written as

$$[\mathrm{d}(\Delta G)/\mathrm{d}T]p,n(i) = -\Delta S, \tag{1}$$

where T is the absolute temperature. ΔG for an electrochemical reaction is related to the cell voltage sustaining the electrochemical reaction by the equation

$$\Delta G = -nFE, \tag{2}$$

where n is the number of electrons transferred and F is the Faraday. Therefore,

$$\Delta S = nF(\mathrm{d}E/\mathrm{d}T)P,n(i) \tag{3}$$

Since

$$\Delta G = \Delta H - T\Delta S, \quad (4)$$

where ΔH is the enthalpy of the reaction,

$$\Delta G = \Delta H - nFT(\mathrm{d}E/\mathrm{d}T). \quad (5)$$

If the temperature coefficient, $\mathrm{d}E/\mathrm{d}T$, is positive, the electrochemical cell will heat on charge and cool on discharge. Lithium-ion batteries have a negative $\mathrm{d}E/\mathrm{d}T$, which means they can overheat during high-current drains [13]. The total heat released, including the reversible thermodynamic heat released along with the irreversible Joule heat from operation of the cell in an irreversible manner, during charge or discharge at a finite current/rate is described by the equation

$$q = T\Delta S + I(E_{\mathrm{OCV}} - E_{\mathrm{T}}) \quad (6)$$

where E_{T} is the terminal voltage and E_{OCV} is the open-circuit voltage. In general, the entropy heat, $T\Delta S$, is negligibly small compared to the irreversible heat, q, which is released when the cell is in operation. Irreversible behavior manifests itself as a departure from the equilibrium or thermodynamic voltage. The total heat released during cell discharge is the sum of the thermodynamic entropy contribution and the irreversible contributions. A proper dissipation of the heat generated is critical to the safety of the cell. Poor heat dissipation can lead to thermal runaway and other catastrophic situations.

3.2. Cell Design

A major factor in the reduced safety of lithium-ion batteries is their design. The temperature of a cell is determined by the balance between the amount of heat generation and dissipation. Obviously, the heat balance depends on the thermal capacity of the cell, as well as the thermal conductivity, emissivity, external surface area and geometry of the cell. At temperatures above 130–150°C, exothermic chemical reactions between the electrodes and electrolyte set in, which further raise the temperature of the cell. Any cell design that cannot dissipate this heat will promote exothermic reactions inside the cell under adiabatic-like conditions, which can rapidly increase the cell's temperature. The increased temperature will further accelerate the chemical reactions and cause even more heat to be produced, eventually resulting in thermal runaway [14–16], the onset temperature of which delineates the safety limits of the device. Any resulting pressure build-up can cause mechanical failures within cells, such as short circuits, premature death of the cell due to irreversible interruptions in the current path, or the distortion, swelling and rupture of cell casing. Processes that trigger thermal runaway include [16,17]: (i) thermal decomposition of the electrolyte, anode and cathode; (ii) reactions of the electrolyte with the charged anode and cathode; and (iii) melting of the separator and the consequent internal short.

3.3. Electrolyte

If there is one determining factor that decides the safety of lithium-ion batteries, it is the electrolyte. Electrolytes must be stable in the electrochemical window within which the cell operates. They must also be thermally stable over a reasonable temperature range, especially when in contact with the active materials. Non-flammability in air at elevated temperatures is another requirement in order to withstand abuse conditions. An electrolyte that meets the above criteria and also possesses desirable properties such as low viscosity and high conductivity still remains a dream. Lithium is intrinsically unstable with any commonly known electrolyte. Today's electrolytes based on alkyl carbonate solvents are known to react vigorously at elevated temperatures with lithiated graphite and delithiated cathodes (e.g., Li_xCoO_2 ($x < 0.5$)) [18–21]. Calorimetric studies have thus become mandatory to determine the safety of electrode–electrolyte combinations. According to Aurbach *et al.* [22], the commonly used electrolytes such as the ones based on $LiPF_6$ in alkyl carbonate solvents are only a compromise: they are flammable and have electrochemical windows of about 4.5V. Although several alternatives such as ionic liquids and alkyl carbonate-based electrolytes containing salts such as lithium bis(oxalato)borate, $LiBC_4O_8$ (LiBOB) [23] and lithium fluoroalkylphosphates (e.g., $Li[PF_3(C_2F_5)_3]$) [24–26] are being considered as substitutes for $LiPF_6$, the immediate solution seems to be additives that can protect electrode-active materials even at high temperatures by forming highly protective films on the electrodes. Such additives must render the electrode-electrolyte interface stable, but also have low flammability with cell venting.

3.4. Active Materials

Commercial lithium-ion batteries are thermally stable up to about 60°C [27], above which their performance declines. Initial reactions are between the anode and the electrolyte [28]. Reactions between the cathode and electrolyte dominate the heat-evolution processes at elevated temperatures [20]. Allowing the latter processes to proceed would lead to disastrous consequences. Violent reactions are known to occur in the charged state of lithium-ion batteries. In fact, at small values of x in Li_xCoO_2, Li_xNiO_2 and $Li_xMn_2O_4$, the cathodes can adversely influence thermal stability [8,29]. Therefore, thermal studies on cathode materials are performed in their delithiated states. Because cell temperatures during abuse reactions can even melt the aluminum current collector used for the positive electrodes, Biensan *et al.* [30] concluded that cell temperatures during cell explosion could shoot above 659°C, the melting point of aluminum. In this Chapter, we discuss safety issues with special reference to the thermal stability of $LiCoO_2$ and $LiCo_{0.2}Ni_{0.8}O_2$ in contact with battery electrolytes.

4. $LiCoO_2$

The most exploited cathode material for commercial lithium-ion batteries is $LiCoO_2$, a layered compound isostructural with the rhombohedral α-$NaFeO_2$ [31]. In its charged state, it is thermally unstable and can decompose, releasing oxygen at temperatures above 200°C [32–36] according to the reaction [8,37]:

$$6\ Li_{0.5}CoO_2 \rightarrow 3\ LiCoO_2 + Co_3O_4 + O_2. \quad (7)$$

The released oxygen can then react with organic solvents to generate heat. ARC studies by Jiang and Dahn [38] showed that organic solvents can reduce $Li_{0.5}CoO_2$ to Co_3O_4 and CoO, eventually even to Co metal. The studies also showed that the reactivity of Li_xCoO_2 with the electrolyte was affected by factors such as particle size, surface area, electrolyte composition, etc. [38,39]. According to MacNeil *et al.* [20], the first thermal processes between Li_xCoO_2 and the electrolyte can be described by an auto-catalytic reaction. In fact, the reaction of $Li_{0.5}CoO_2$ with EC–DEC begins at 130°C, which is much lower than the decomposition temperature of $Li_{0.5}CoO_2$ itself [32]. Jiang and Dahn [38] showed that the reactivity of $Li_{0.5}CoO_2$ was higher in LiBOB–EC–DEC than in $LiPF_6$–EC–DEC. Although LiBOB can effectively stabilize the SEI of LiC_6, the lower stability of the LiBOB-based electrolyte with respect to the cathode would mean that graphite/$LiCoO_2$ cells cannot be rendered safer by replacing $LiPF_6$ with LiBOB in the electrolyte. Baba *et al.* [35], who evaluated the thermal stability of chemically delithiated $LiCoO_2$ ($Li_{0.49}CoO_2$) by differential scanning calorimetry (DSC), the decomposition of EC ($C_3H_4O_3$) by the cathode material begins at 190°C according to the reaction [32]:

$$10\ Li_{0.5}CoO_2 + C_3H_4O_3 \rightarrow 5\ LiCoO_2 + 5\ CoO + 3\ CO_2 + 2\ H_2O. \quad (8)$$

The DSC pattern also showed a peak at 230°C, which Baba *et al.* [35] ascribed to the oxidation of the electrolyte caused by oxygen released from $Li_{0.49}CoO_2$.

4.1. Coated $LiCoO_2$ Cathodes

It is recognized that many safety-related incidents are due to vigorous exothermic reactions between delithiated cathodes and electrolytes [32,35,40]. Naturally, a logical approach to improve thermal stability would be to provide a barrier that prevents direct reaction between the constituents. Such barriers can be effected by modifying cathode surfaces with inorganic coating or by adding film-forming additives to the electrolyte [41–45]. For example, γ-butyrolactone as an additive in the electrolyte has been reported to decompose and form products that encapsulate the cathode [46]. In fact, lithium-ion cells with this additive did not explode during nail penetration tests at 4.35V and overcharge tests up to 12V. However, concerns about the compatibility of γ-butyrolactone with the anode and cathode remain.

Lee *et al.* [47] used a coating of gel polymer electrolyte based on cPVA (cyano-substituted polyvinyl alcohol). The –CN group gives the polymer a high dielectric constant (ε = 15 at 1 kHz at 20°C), which facilitates dissociation of lithium salts, leading to high ionic conductivity (around 7 $mS.cm^{-1}$ at 25°C). Figure 1 shows the DSC profiles of the bare and cPVA-coated $LiCoO_2$ charged to 4.2 V. The bare $LiCoO_2$ exhibits large exothermic peaks ($\Delta H = 413\ J.g^{-1}$) between 100 and 300°C [32,35,40]. However, the cPVA-coated $LiCoO_2$, showed a noticeable decrease in exothermic heat ($\Delta H = 31\ J.g^{-1}$), suggesting that the polymer electrolyte encapsulates the cathode particles, effectively suppressing the exothermic reaction [47]. The authors [47] proposed that the –CN moiety coordinates with the cobalt cation in $LiCoO_2$, imparting structural stability to the delithiated cathode, as well as significantly suppressing exothermic reactions.Recently, several papers reported enhancement in the cyclability of cathodes by coating cathode particles with oxides, glasses, etc. [48–58]. A benefit of coated electrodes is an increase in their thermal stability during contact with the electrolyte. Cho *et al.* [59] showed that $LiCoO_2$ coated with nanoparticulate $AlPO_4$ blocked the thermal runaway of lithium-ion cells, in addition to significantly reducing electrolyte oxidation and cobalt dissolution into the electrolyte. In another study, Cho *et al.* [60] demonstrated the 12 V overcharge behavior of the $AlPO_4$-coated $LiCoO_2$ in terms of its exothermic behavior. Cho [61] also showed that an $AlPO_4$ coating thickness of 20 nm was optimal. Although increased thickness of the coating drastically reduced the exothermic reaction between

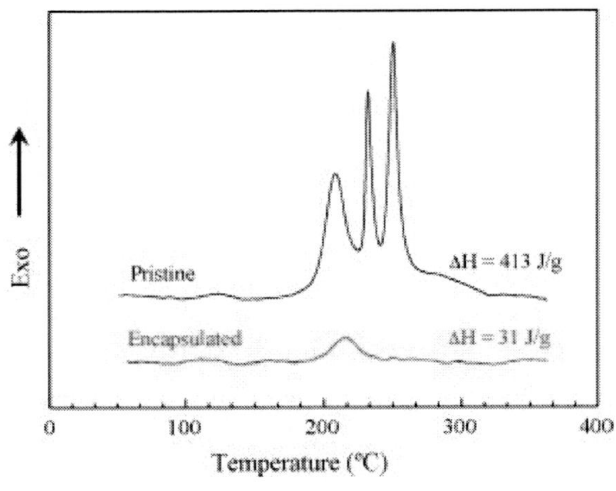

Figure 1. DSC profiles of the bare and cPVA-coated delithiated $LiCoO_2$ active materials [47].

Li_xCoO_2 and the electrolyte in addition to increasing the onset temperature of oxygen evolution, the reduced lithium-ion diffusivity was detrimental to cycling performance. Fey *et al.* [62] showed that $LiCoO_2$ coated with cobalt oxides displayed increased resistance to decomposition reactions with the electrolyte. Not only was the temperature of the reaction raised by 11◦C, but the coating also reduced the exothermicity of the reaction (Figure 2).

Although the charge decomposition of $LiCoO_2$ ($Li_{0.5}CoO_2$) takes place above 200°C in the absence of an electrolyte (according to equation 7) [32], its decomposition temperature is lowered to 130°C in the presence of EC:DEC [63]. In line with the fact that the reactivity of $LiCoO_2$ would increase with increased levels of delithiation, Fey *et al.* [63] found that the onset temperatures (T_o) for decomposition were 151°C and 110°C for $LiCoO_2$ charged to 4.2 and 4.5 V, respectively (Table I). The corresponding values for $LiCoO_2$ coated with $MgAlO_4$ were 175°C and 145°C, suggesting that thermal stability improved due to the coating. Moreover, the greatest areas of the DSC peaks of the charged $LiCoO_2$ electrodes were larger than those for the charged coated electrodes (Figure 3), which indicates that the amount of heat generated (a measure of the amount of oxygen released) was reduced upon coating. The lower exothermicity and higher onset temperature are proof of the higher thermal stability of the coated electrodes.

Fey *et al.* [63] also demonstrated the higher thermal stability of $LiCoO_2$ coated with lanthanum aluminum garnet ($La_3Al_5O_{12}$) as well as with $Li_4Ti_5O_{12}$ and $Li_4Mn_5O_{12}$. For example, their DSC studies showed that coatings with $Li_4Ti_5O_{12}$ and $Li_4Mn_5O_{12}$ raised the onset temperature of a $LiCoO_2$ sample charged to 4.2 V by 7°C and 5°C to 471°C and 469°C, respectively. The coated samples also reduced the exothermic heat from 131 $J.g^{-1}$ to 120 $J.g^{-1}$ and 122 $J.g^{-1}$, respectively, for $Li_4Ti_5O_{12}$ and $Li_4Mn_5O_{12}$ coatings.

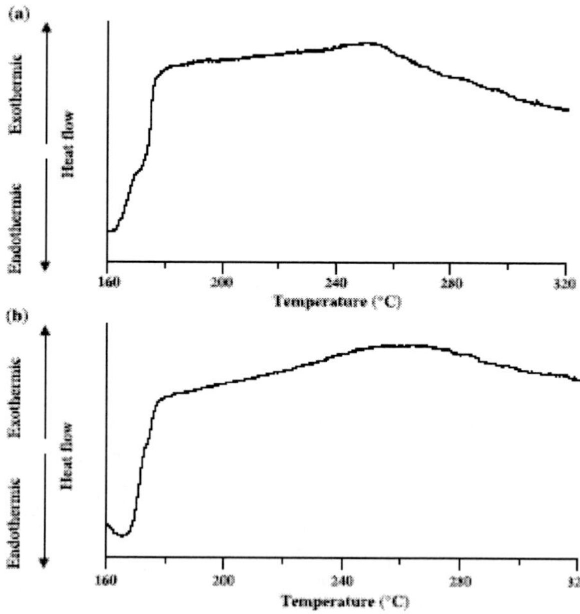

Figure 2. DSC profiles of (a) bare and (b) 0.3-wt.% cobalt oxide-coated $LiCoO_2$ [62].

Table I. Data derived from DSC profiles (figure 3) for bare and $MgAlO_4$-coated $LiCoO_2$ charged to various voltages.

Voltage	Cathode Material	T_o (°C)[a]	T_d (°C)[b]	H_{tot} (J g^{-1})[c]
4.5 V	Bare	110	174	205
	Coated	145	178	182
4.4 V	Bare	120	182	185
	Coated	155	188	155
4.3 V	Bare	148	186	164
	Coated	156	190	142
4.2 V	Bare	151	191	131
	Coated	175	201	121

[a] Onset temperature; [b] decomposition temperature; [c] total evolved heat.

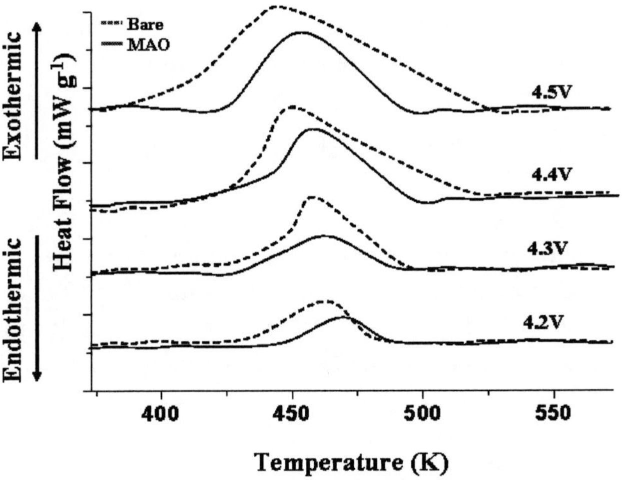

Figure 3. DSC profiles of bare $LiCoO_2$ and $MgAlO_4$-coated $LiCoO_2$ charged to various voltages.

5. $LiCo_{0.2}Ni_{0.8}O_2$

The general consensus is that solid solutions of the general formula $LiNi_{1-y}Co_yO_2$ are structurally more stable than their pristine end-member homologues and also exhibit superior performance [64–67]. Typically, the solid solution of the formula $LiNi_{0.8}Co_{0.2}O_2$, with its higher reversible capacity than $LiCoO_2$, as well as significant cost and performance benefits, is considered a potential next-generation cathode material [68,69]. Several groups have reported the formation of surface films on cathodes such as $LiCoO_2$, $LiMn_2O_4$, and $LiNi_{1-x}Co_xO_2$ [70–78], but the nature of the reactions that generate the films is unclear. According to Abraham *et al.* [79], an oxygen-deficient surface layer was formed on $LiNi_{1-x}Co_xO_2$ as a result of oxygen transfer reactions with the electrolyte. The thermal characteristics of $LiNi_{1-x}Co_xO_2$ is reminiscent of those of $LiNiO_2$, which in its delithiated state has a poor thermal stability due to the presence of the unstable Ni^{4+} ion [80]. At an x value of 0.3 in Li_xNiO_2, the compound releases oxygen at a lower temperature than Li_xCoO_2 (x = 0.4). The lower stability of Li_xNiO_2 is attributed to the easier reduction of Ni^{3+} compared to Co^{3+} [81]. Therefore, a cell with $LiNiO_2$ should be less tolerant than one with $LiCoO_2$ under abusive conditions [8]. Ohzuku *et al.* [82] showed that Li_xNiO_2 ($x = 0.15$) undergoes an exothermic reaction at about 200◦C. Arai *et al.* [81] reported that Li_xNiO_2 transforms into a rocksalt structure at 200◦C due to cation mixing. In the absence of air, $LiNiO_2$ decomposes, releasing oxygen [83]:

$$4\ LiNiO_2 \rightarrow 2\ Li_2O + 4\ NiO + O_2. \qquad (9)$$

Below x values of 0.25, Li_xNiO_2 undergoes highly exothermic reactions with common electrolytes with an onset temperature around 200◦C. Reflecting the lower thermal stability of $LiNiO_2$, Fey *et al.* [63] found that major exothermic reactions of $LiNi_{0.8}Co_{0.2}O_2$ charged to 4.5V set in around 168°C. However, this value is lower than the 222°C reported by Ha *et al.* [84].

5.1. Substituted $LiNi_yCo_{1-y}O_2$ Compositions

Several substituted compositions based on $LiNi_yCo_{1-y}O_2$ have been investigated with the dual goal of improving cycle life and thermal stability. A composition such as $LiNi_{0.8}Co_{0.15}Al_{0.05}O_2$ has been reported to have poor cycle life, attributable to greater impedance at the cathode [85,86]. The rise in impedance has been traced to the presence of a NiO-type interfacial film formed by the reduction of the electrolyte by highly oxidizing and unstable Ni^{4+} generated during the charge process [87]. A way out to bypass this rise is to opt for a similar but more stable cathode material such as $LiNi_{1/3}Co_{1/3}Mn_{1/3}O_2$ [88,89].

5.2. Coated $LiNi_yCo_{1-y}O_2$ Compositions

Oxygen release at elevated temperatures is believed to trigger electrolyte decomposition mainly through surface reactions [35]. Therefore, it is logical to expect that coatings would suppress them. A composition of interest with a higher specific capacity than $LiCoO_2$ is $LiNi_{0.8}Co_{0.1}Mn_{0.1}O_2$. However, its practical application is hindered by its relatively poor thermal stability. In order to improve its thermal characteristics, Cho *et al.* [90] coated the cathode material with $AlPO_4$. The coating did not affect the specific capacity (188 mAh/g at a cut-off voltage of 4.3 V for the bare compound), but noticeably suppressed violent exothermic reactions between the cathode and the electrolyte. Safety tests conducted on cells with the $AlPO_4$-coated $LiNi_{0.8}Co_{0.1}Mn_{0.1}O_2$ cathode exhibited excellent overcharge performance. Moreover, they did not experience thermal runaway, smoking or explosion, in contrast to those containing the uncoated cathode material [90]. In a comparative study of cells with $AlPO_4$-coated Li_xCoO_2 and $AlPO_4$-coated $Li_xNi_{0.8}Co_{0.1}Mn_{0.1}O_2$, Cho *et al.* [62] found that as the rate was increased from 1C to 3 C, the surface temperature of the cell with the coated $Li_xNi_{0.8}Co_{0.1}Mn_{0.1}O_2$ did not exceed 125◦C, while that of the cell with the coated Li_xCoO_2 exceeded 170◦C.

Recently, Fet *et al.* [63] carried out a detailed study of the thermal behavior of $LiNi_{0.8}Co_{0.2}O_2$ coated with $MgAlO_4$. Figure 4 gives the DSC profiles of bare and $MgAlO_4$-coated $LiNi_{0.8}Co_{0.2}O_2$. Major exothermic reactions of the bare cathode material charged to 4.5V occur around 168C. It can be seen from Table II that the charged coated material exhibited not only higher T_d and T_o than the bare sample, but lower H_{tot} too, suggesting the higher stability of the coated material. They also investigated the thermal behavior of charged $LiNi_{0.8}Co_{0.2}O_2$ coated with double oxides such as $La_3Al_5O_{12}$, $Li_4Ti_5O_{12}$ and $Li_4Mn_5O_{12}$ [63]. Their DSC data showed a higher decomposition temperature for the coated electrodes, suggesting a reduction in oxygen generation upon coating [91].

6. Conclusions

Commercial lithium-ion batteries based on flammable non-aqueous electrolytes and layered oxides such as $LiCoO_2$ and $LiNi_{0.8}Co_{0.2}O_2$ are inherently unsafe. Thermal reactions in such cells can be triggered by the oxidation of solvents by the oxygen released by the unstable charged cathodes during thermal events. Therefore, it appears that there is probably no possibility to suppress such hazards with cathodes, although limited success has been achieved by using coated cathodes and electrolyte additives that react with the cathode surface to generate a barrier between the active material and the electrolyte. With such cathode-active materials, substituents that can push the thermal decomposition point of the material to above 300°C without compromising capacity are needed, but in all likelihood, cathodes that are not prone to release oxygen, such as $LiFePO_4$, are probably the only practical solution for thermally safe cathodes.

Table II. Data derived from DSC profiles (figure 4) for bare and $MgAlO_4$-coated $LiCoO_2$ charged to various voltages.

Voltage	Cathode Material	T_o (°C)[a]	T_d (°C)[b]	H_{tot} (J g^{-1})[c]
4.5 V	Bare	109	168	230
	Coated	145	186	195
4.4 V	Bare	118	175	210
	Coated	172	214	173
4.3 V	Bare	120	195	190
	Coated	181	227	155
4.2 V	Bare	148	200	180
	Coated	221	268	150

[a] Onset temperature; [b] decomposition temperature; [c] total evolved heat.

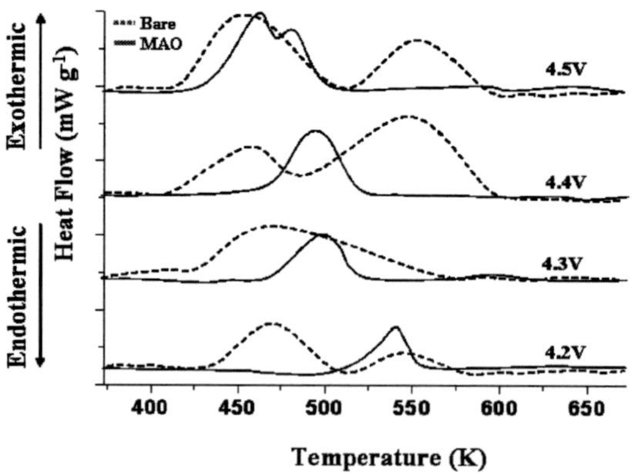

Figure 4. DSC profiles of bare $LiNi_{0.8}Co_{0.2}O_2$ and $MgAlO_4$-coated $LiNi_{0.8}Co_{0.2}O_2$ charged to various voltages.

REFERENCES

[1] Balakrishnan, PG; Ramesh, R; Prem Kumar, T. *J. Power Sources.*, 2006, *155*, 401.
[2] Tarascon, JM; McKinnon, WR; Coowar, F; Bowmer, TN; Amatucci, GG; Guyomard, D. *J. Electrochem. Soc.*, 1994, 141, 1421.
[3] Amatucci, GG; Tarascon, JM; Klein, LC. *Solid State Ionics.,* 1996, 83, 167.
[4] Jang, DH; Shin, YJ; Oh, SM. *J. Electrochem. Soc.*, 1996, 143, 2204.
[5] Tarascon, JM; Guyomard, D; Baker, GL. *J. Power Sources.,* 1993, 43–44, 689.
[6] Guyomard, D; Tarascon, JM. *J. Electrochem. Soc.*, 1993, 140, 3071.
[7] Tarascon, JM; Guyomard, D. *Electrochim. Acta.,* 1993, 38, 1221.
[8] Dahn, JR; Fuller, EW; Obrovac, M; von Sacken, U. *Solid State Ionics.,* 1994, 69, 265.
[9] Tarascon, JM; Guyomard, D. *Solid State Ionics.,* 1994, 69, 293.
[10] Guyomard, D; Tarascon, JM. *J. Power Sources.,* 1995, 54, 92.
[11] Chen, Y; Evans, JW. *J. Electrochem. Soc.*, 1996, 143, 2708.
[12] Surianarayanan, M; Vijayaraghavan, R; Swaminathan, G; Rao, PG. *Curr. Sci.*, 2001, 80, 738.
[13] Shukla, AK; Prem Kumar, T. *Curr. Sci.*, 2008, 94, 314.
[14] von Sacken, U; Nodwell, E; Sundher, A; Dahn, JR. *Solid State Ionics.,* 1994, 69, 284.
[15] Levy, SC; Bro, P. *Battery Hazards and Accident Prevention*, Plenum, New York, 1994.
[16] Tobishima, S; Yamaki, JI. *J. Power Sources.,* 1999, 81–82, 882.
[17] Yamaki, JI. In Lithium Ion Batteries, Wakihara, M; Yamamoto, O. (Eds.) Wiley/VCH/Kodansha, Tokyo, 1998, 83.
[18] Richard, MN; Dahn, JR. *J. Electrochem. Soc.*, 1999, 146, 2068.
[19] MacNeil, DD; Larcher, D; Dahn, JR. *J. Electrochem. Soc.*, 1999, 146, 3596.
[20] MacNeil, DD; Christensen, L; Landucci, J; Paulsen, JM; Dahn, JR. *J. Electrochem. Soc.*, 2000, 147, 970.
[21] MacNeil, DD; Dahn, JR. *J. Phys. Chem.*, 2001, A105, 4430.
[22] Aurbach, D; Talyosef, Y; Markovsky, B; Markevich, E; Zinigrad, E; Asraf, L; Gnanaraj, JS; Kim, HJ. *Electrochim. Acta.,* 2004, 50, 247.
[23] Xu, K; Zhang, SS; Jow, TR; Xu, W; Angell, CA. *Electrochem. Solid-State Lett.*, 2002, 5, A26.
[24] Schmidt, M; Heider, U; Kuehner, A; Oesten, R; Jungnitz, M; Ignatev, N; Sartori, P. *J. Power Sources.,* 2001, 97–98, 557.

[25] Oesten, R; Heider, U; Schmidt, M. *Solid State Ionics.,* 2002, 148, 391.

[26] Gnanaraj, JS; Levi, MD; Gofer, Y; Aurbach, D. *J. Electrochem. Soc.*, 2003, 150, 445.
[27] Zhang, SS; Xu, K; Jow, TR. *Electrochem. Solid-State Lett.*, 2002, 5, A206.
[28] MacNeil, DD; Lu, Z; Chen, Z; Dahn, JR. *J. Power Sources.,* 2002, 108, 8.
[29] Zhang, Z; Fouchard, D; Rea, JR. *J. Power Sources.,* 1998, 70, 16.
[30] Biensan, Ph; Simon, B; Peres, JP; de Guibert, A; Broussely, M; Bodet, JM; Perton, F. *J. Power Sources.,* 1999, 81–82, 906.
[31] Dyer, LD; Borie, Jr., KBS; Smith, GP; *J. Am. Chem. Soc.*, 1954, 76, 1499.
[32] MacNeil, DD; Dahn, JR. *J. Electrochem. Soc.*, 2001, 148, A1205.
[33] Cho, J; Kim, YJ; Kim, TJ; Park, B. *Chem. Mater.*, 2001, 13, 18.
[34] MacNeil, DD; Dahn, JR. *J. Electrochem. Soc.*, 2002, 149, A912.
[35] Baba, Y; Okada, S; Yamaki, JI. *Solid State Ionics.,* 2002, 148, 311.
[36] Yamaki, JI; Baba, Y; Katayama, N; Takatsuji, H; Egashira, M; Okada, S. *J. Power Sources.,* 2003, 119–121, 789.
[37] Kweon, HJ; Kim, SJ; Park, DG. *J. Power Sources.,* 2000, 88, 255.
[38] Jiang, J; Dahn, JR. *Electrochim. Acta.,* 2004, 49, 2661.
[39] MacNeil, DD; Dahn, JR. *J. Electrochem. Soc.*, 2003, 151, A21.
[40] Leising, RA; Palazzo, MJ; Takeuchi, ES; Takeuchi, KJ. *J. Electrochem. Soc.*, 2001, 148, A838.
[41] MacNeil, DD; Dahn, JR. *J. Electrochem. Soc.*, 2001, 148, A1211.
[42] Cho, J. *Electrochem. Commun.*, 2003, 5, 146.
[43] Cho, J; Kim, Y; Kim, B; Lee, J; Park, B. *Angew. Chem. Int. Ed.*, 2003, 42, 1618.
[44] Kawamura, T; Kimura, A; Egashira, M; Okada, S; Yamaki, JI. *J. Power Sources.,* 2002, 104, 260.
[45] Zhang, SS; Xu, K; Jow, TR. *J. Power Sources.,* 2003, 113, 166.
[46] Takami, N; Inagaki, H; Ueno, R; Kanda, M. 11th International Meeting on Lithium Batteries, Monterey, CA, June 23–28, 2002.
[47] Lee, SY; Kim, SK; Ahn, S. *J. Power Sources.,* 2007, 174, 480.
[48] Kuribayashi, I; Yokoyama, M; Yamashita, M. *J. Power Sources.,* 1995, 54, 1.
[49] Yoshio, M; Wang, H; Fukuda, K; Hara, Y; Adachi, Y. *J. Electrochem. Soc.*, 2000, 147, 1245.
[50] Yu, P; Ritter, JA; White, RE; Popov, BN. *J. Electrochem. Soc.*, 2000, 147, 1280.
[51] Hossain, S; Loutfy, R; Saleh, Y; Kim, Y. *Proc. of the 40th Power Sources Conference*, Cherry Hill, USA, June 2002, 29.
[52] Cho, J; Kim, YJ; Park, B. *Chem. Mater.*, 2000, 12, 3788.
[53] Cho, J; Kim, YJ; Park, B. *J. Electrochem. Soc.*, 2001, 148, A1110.
[54] Mladenov, M; Stoyanova, R; Zhecheva, E; Vassilev, S. *Electrochem.*

Commun. 2001, 3, 410.
[55] Wang, Z; Wu, C; Liu, L; Wu, F; Chen, L; Huang, X. *J. Electrochem. Soc*., 2002, 149, A466.
[56] Kannan, AM; Rabenberg, L; Manthiram, A. *Electrochem. Solid-State Lett*., 2003, 6, A16.
[57] Fey, GTK; Yang, HZ; Prem Kumar, T; Naik, SP; Chiang, AST; Lee, DC; Lin, JR. *J. Power Sources.,* 2004, 132, 172.
[58] Fey, GTK; Weng, ZX; Chen, JG; Lu, CZ; Prem Kumar, T; Naik, SP; Chiang, AST; Lee, DC; Lin, JR. *J. Appl. Electrochem*., 2004, 34, 715.
[59] Cho, J; Kim, YJ; Kim, TJ; Park, B. *Angew. Chem. Int. Ed*., 2001, 40, 3367.
[60] Cho, J; Kim, H; Park, B. *J. Electrochem. Soc*., 2004, 151, A1707.
[61] Cho, J. *Electrochim. Acta.,* 2003, 48, 2807.
[62] Fey, GT. K; Lin, YY; Prem Kumar, T. *Surf. Coat. Technol*., 2005, 191, 68.
[63] Fey, GTK; Chen, YK; Chen, YG; Chang, BF. Unpublished results. Suresh, P; Rodrigues, S; Shukla, AK; Sivashankar, SA; Munichandraiah, N. *J. Power Sources.,* 2002, 112, 665.
[64] Fey, GTK; Weng, ZX; Chen, JG; Prem Kumar, T. *Mater. Chem. Phys*., 2003, 82, 5.
[65] Fey, GTK; Chen, JG; Wang, ZF; Yang, HZ; Prem Kumar, T. *Mater. Chem. Phys*., 2004, 87, 246.
[66] Elumalai, P; Vasan, HN; Munichandraiah, N. *Mater. Res. Bull*., 2004, 39, 1895.
[67] Delmas, C; Saadoune, I; Rougier, A. *J. Power Sources.,* 1993, 43–44, 595.
[68] Aragane, J; Matsui, K; Andoh, H; Suzuki, S; Fukuda, H; Ikeya, H; Kitaba, K; Ishikawa, R. *J. Power Sources.,* 1997, 68, 13.
[69] Wang, Z; Huang, X; Chen, L. *J. Electrochem. Soc*., 2003, 150, A199.
[70] Aurbach, D. *J. Power Sources.,* 1999, 81–82, 95.
[71] Aurbach, D. *J. Power Sources.,* 2000, 89, 206.
[72] Balasubramanian, M; Lee, HS; Sun, X; Yang, XQ; Moodenbaugh, AR; McBreen, J; Fischer, DA; Fu, Z. *Electrochem. Solid-State Lett*., 2002, 5, A22.
[73] Wang, Y; Guo, X; Greenbaum, S; Liu, J; Amine, K. *Electrochem. Solid-State Lett*., 2002, 4, A68.
[74] Andersson, AM; Abraham, DP; Haasch, R; MacLaren, S; Liu, J; Amine, K. *J. Electrochem. Soc*., 2002, 149, A1358.
[75] Eriksson, T; Andersson, AM; Bishop, AG; Gejke, C; Gustafsson, T; Thomas, TO; *J. Electrochem. Soc*., 2002, 149, A69.
[76] Vogdanis, L; Heitz, W. *Macromol. Rapid Commun*., 1986, 7, 543.
[77] Vogdanis, L; Martens, B; Uchtmann, H; Hensel, F; Heitz, F. *Macromol. Chem*., 1990, 191, 465.

[78] Abraham, DP; Twesten, RD; Balasubramanian, M; Kropf, J; Fischer, D; McBreen, J; Petrov, I; Amine, K. *J. Electrochem. Soc*., 2003, 150 , A1450.
[79] Ohzuku, T; Yanagawa, T; Kougushi, M; Ueda, A. *J. Power Sources.,* 1997, 68, 131.
[80] Arai, H; Okada, S; Sakurai, Y; Yamaki, JI. *Solid State Ionics.,* 1998, 109, 295.
[81] Ohzuku, T; Ueda, A; Kouguchi, M. *J. Electrochem. Soc*., 1995, 142, 4033.
[82] Li, W; Currie, JC; Wolstenholme, J. *J. Power Sources.,* 1997, 68, 565.
[83] Ha, HW; Jeong, KH; Yun, NJ; Hong, MZ; Kim, K. *Electrochim. Acta.,* 2005, 50, 3764.
[84] Zhang, X; Ross Jr., PN; Kostecki, R; Kong, F; Sloop, S; Kerr, JB; Striebel, K; Cairns, EJ; McLarnon, F. *J. Electrochem. Soc*., 2001, 148, A463.
[85] Balasubramanian, M; Sun, X; Yang, XQ; McBreen, J. *J. Power Sources.,* 2001, 92, 1.
[86] Abraham, DP; Twesten, RD; Balasubramanian, M; Petrov, I; McBreen, J; Amine, K. *Electrochem. Commun*., 2002, 4, 620.
[87] Belharouak, I; Sun, YK; Liu, J; Amine, K. *J. Power Sources.,* 2003, 123, 247.
[88] Belharouak, I; Lu, WQ; Vissers, D; Amine, K. *Electrochem. Commun*., 2006, 8, 329.
[89] Cho, J; Kim, TJ; Kim, J; Noh, M; Park, B. *J. Electrochem. Soc*., 2004, 151, A1899.
[90] Belharouak, I; Tsukamoto, H; Amine, K. *J. Power Sources.,* 2003, 119, 175.

In: Lithium Batteries: Research, Technology… ISBN: 978-1-60741-722-4
Editors: Greger R. Dahlin, et al.

Chapter 6

COMPOSITIONAL AND STRUCTURAL EVOLUTION OF CATHODE PARTICLES OF THE CYCLED LITHIUM BATTERIES INVESTIGATED BY ANALYTICAL HIGH RESOLUTION TRANSMISSION ELECTRON MICROSCOPY(AHRTEM)

Yuewu Zeng [1*], Shaofeng Chen[2*], Jinhua He[2] and Z.C. Kang [2,3*]

[1] Center of analysis and measurement, Zhejiang University, Hangzhou 310028, China

[2] Ningbo Jinhe New Materials Inc, Zhejiang Yuyao 315400 China

[3] RE Power-Tech 8211.E.Garfield St. J-201 Scottsdale AZ 85257, USA

[*] Corresponding authors: E-mail: ywzeng@ema.zju.edu.cn

[*] E-mail: ceo@nicosn.com

[*] E-mail: zckang@repowertech.com

1. INTRODUCTION

1.1 The Cathode of Lithium Battery is the Li^+ Source and Sinks

As is well known [1-3], the lithium battery is a rechargeable battery and its lithium comes from the cathode electrode materials such as lithium intercalated transition metal oxides, for example, $LiCoO_2$, $LiNiO_2$, $LiMnO_2$, $Li(Co_{1-x-y}Ni_xMn_y)O_2$, and $LiMn_2O_4$. During the charging process, the Li^+ ions pull out from the lithium intercalated oxides by electric field and are expelled into the carbon layers of graphite anode through an electrolyte. Therefore, the graphite anode acts as Li^+ ions sink. However, during the discharging process, the Li^+ ions stored in the graphite anode act as Li+ source and will flow out from the graphite anode intercalating into the oxygen closed packed layers of the dioxide cathode through an electrolyte. So, the cathode is a sink of the flowing Li^+ ions. The anode and cathode both act as Li^+ ion source and sink. The capacity of the lithium battery is dominated by the Li^+ ion source storage capacity and the sink volume. The rate of Li^+ ions flow is also related to the source and sink capability. The cathode and anode, especially the cathode, are very important for the lithium battery.

The Li^+ ions source and sink of a lithium battery are crystalline compounds: oxides and graphite. Li^+ ions inserting or extracting from the compounds have to make these compounds to capture (during inserting) or to release (if extracting) electron, which means a redox process has to occur in these crystals. It was known that Fe, Co, Ni, and Mn can form M_2O_3 and M_3O_4, in which the oxygen assembles a close-packing array and the M occupy the octahedron and/or tetrahedron voids based on the ratio of metal and oxygen. Figure 1a shows the location of metal atoms of FeO, Fe_2O_3 and Fe_3O_4 on the oxygen closed packing layer. The Fe^{3+} occupy 2/3 octahedron voids of a oxygen closed packed layer in Fe_2O_3 to create a Corundum structure. The Fe^{2+} and Fe^{3+} cations occupy alternately at 1/2 tetrahedron voids and all of the octahedron voids forming Spinel structure. If Fe^{2+} cation located at all of the octahedron voids of the oxygen closed packed layer, it will create a FeO with NaCl structure (face center cubic (f.c.c. lattice)). Usually Fe atom have 2+ and 3+ valence state and Fe^{3+} prefer to locate at octahedron, but Fe^{2+} favor being in the tetrahedron. However, if Na^+ involved in FeO structure and the Na and Fe ordered as they alternately located in different octahedron layers, then $NaFeO_2$ oxide is formed as shown in Figure 1b. $NaFeO_2$ can be synthesized by hydrothermal method. Using Li to substitute Na in the $NaFeO_2$ can form

$LiFeO_2$, but it is not stable. However, $LiCoO_2$ and $LiNiO_2$ can be synthesized by different methods, but they always have $NaFeO_2$–related structure.

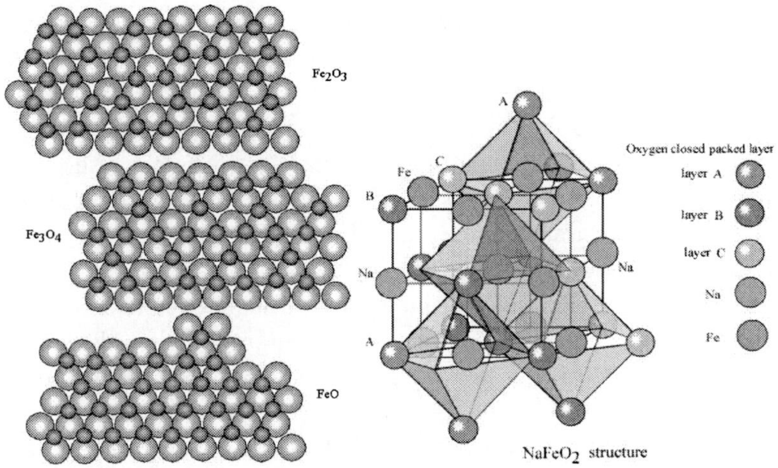

Figure 1. (a) Oxygen closed packed array and Fe atom located at octahedron and tetrahedron voids for FeO, Fe_2O_3 and Fe_3O_4 (b) $NaFeO_2$ structure

During charging or discharging process, the lithium content of $LiMO_2$ (M=Co, Ni, Mn) cathode has to vary from 1 to x (0<x<1), that implies the oxide is nonstoichiometric and the metal has different ratio of M^{3+} and M^{4+}. The voltage of the cell related to the transference number, t_{Li+} , of Li^+ ions and lithium content, x, of the cathode, Li_xMO_2, may determine the chemical potential of the oxide. Then

$$V=1/4F\ t_{Li+}\ d\mu = \mu_2-\mu_1/4F = \mu^o{}_2-\mu^o{}_1/4F - RT/4F\ \text{In}\ a_c(x)/a_a(1-x)$$

$$=V^o - RT/4F\ \text{In}\ a_c(x)/a_a(1-x)$$

$$= [E_f\ (Li^+) - E_f(M^{3+}/M^{4+})] - RT/4F\ \text{In}\ a_c(x)/a_a(1-x) \qquad (1\text{-}1)$$

Where: $E_f(Li^+)$ is Fermi energy of the Li metal and $E_f(M^{3+}/M^{4+})$ is the Fermi energy of M^{3+}/M^{4+} redox couple; x is lithium content of cathode and (1-x) is lithium content of anode; F is Faraday's constant. It is clear that the voltage of the Lithium battery is related to the electron band structure, which is dominated by the crystal structure and the voltage decreasing is related to the chemical activity of Li^+ intercalated layer metal oxides (cathode) and graphite (anode). Nonstoichiometry and phase transitions of the nonstoichiometric Li+ transition metal dioxides should dominate the electrochemical property of Li^+ battery.

In crystalline solid, the electric neutralization should be held at any place of the solid. For the cathode and anode of Li+ battery, the motion of the most mobile ions generally determines the rate of the charging and discharging process. However, the ionic transport rate may be largely influenced by interaction with the electrons and holes of the electrodes because the electric neutralization has to be obeyed. Wagner[4] indicated that the ionic diffusivity, D_i, and the electron transference number, t_e, chemical activity of the ion species, and concentration of the ions in the solid dominate the ionic flows. Therefore, wide nonstoichiometry and electron transference number close to 1 will have more advantageous. Due to Li^+ ion migration should be guided by the navigation of electrons or polarized electron cloud. So the cathode having electron and ionic mixed conductivity would be able to enhance the charging and discharging process. It was known [4] that high concentration of mobile ion and low concentration of electron in a cathode may have higher mobility of Li^+ ions. The range and phase of the Li^+ nonstoichiometry of the crystalline cathode could determine its electric and ionic conductivity resulting excellent or poor electrochemical performance of a Li^+ battery.

Based on our knowledge of the Li^+ battery we believe that revealing the structure variation of the cathode crystalline particles of a discharged or charged Li^+ battery is great important for development of high performance Li^+ battery. Therefore, we used snapshot method, which take the cathode particle underwent 300 discharge and charge cycles, to observe the surface and structure of the particles at atomic scale by AHRTEM (Analytic High Resolution transmission Electron Microscopy). In this paper we review the compositional and structural evolution of the cycled cathode particles of the Li^+ battery.

1.2 The Compositional and Structural Feature of Surface of a Cathode Particle

$LiMO_2$(M=Co, Ni, Mn) have $NaFeO_2$-type structure which means the Li and M (Co, Ni, Mn) alternately distribute in the oxygen closed packed layers, which are the {111} planes of a face center cubic lattice as shown in figure 1b. The crystallographic unit cell of $LiMO_2$ (for example $LiCoO_2$) is hexagonal (R-3) as shown in figure 2. The basic plane of the hexagonal unit cell is a

{111} plane of $NaFeO_2$ and the Li occupy the octahedron void created by the successive oxygen closed packed layers (in figure 2 it marked as O_A,O_B) and the Co located in other octahedron site formed by other successive oxygen closed packed layers (in figure 2 it marked as O_B and O_C). Therefore the surface $(0001)_{hex}$ is hard to pass through Li ions, but Li+ ions are easy to escape from the $\{1\text{-}100\}_{hex}$ or $\{1\text{-}210\}_{hex}$ surfaces of the $LiCoO_2$. Therefore, The cathode particle exposing these surfaces to electrolyte should make the Li+ ions to intercalate or de-intercalate more easy and also can benefit the charge or discharge process.

1.3 Fundamental Structural and Compositional Relationships between the $NaFeO_2$ and $LiMO_2$ (M=Co,Ni,Mn)

As mentioned before, the Li^+ inserted into the successive oxygen closed packed layers will reduce the transition metal cation:

$$M^{4+} + e^- = M^{3+}$$

This process may cause following sub-reaction $2M^{3+} \rightarrow M^{4+} + M^{2+}$

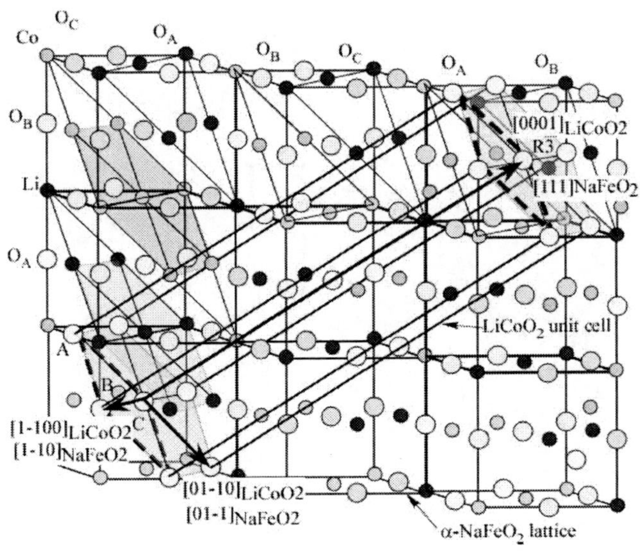

Figure 2. The crystallographic unit cell (Hexagonal) of $LiCoO_2$ and $NaFeO_2$ frame (f.c.c lattice).

It called disproportion. Different cation may prefer to occupy different coordination environment. For example Ni^{2+} favor at tetrahedron site. Therefore the cathode materials, $LiCoO_2$, $Li(Co_{1-x}Ni_x)O_2$, $Li(Co_{1/3}Ni_{1/3}Mn_{1/3})O_2$ and $Li(Ni_{1-x}Mn_x)O_2$ may have ordered cations and/or ordered Li ions arrays. But if the Co, Ni, and Mn still located in same successive oxygen closed packed layers, or in other words stay in same octahedron slab, then the ordered structure and $NaFeO_2$ sub-lattice would hold same relationship as $LiCoO_2$. For example the ordered $Li(Ni_{1/3}Co_{1/3}Mn_{1/3})O_2$ unit cell shown in figure 3.

Li ion usually occupies at the octahedron site, but it also may stay in a tetrahedron site and the Ni^{2+} also is favor to occupy at tetrahedron site. Therefore if the disproportionate reaction is occurred and Ni^{2+} can be formed, then the formed Ni^{2+} may move to the tetrahedron site created by successive octahedra shown in figure 4. The Ni^{2+} or Li^+ located at tetrahedron will increase the barriers of migrated Li+ ions and induce structural change from hexagonal to spinel structure as Fe_3O_4 shown in figure 1a.

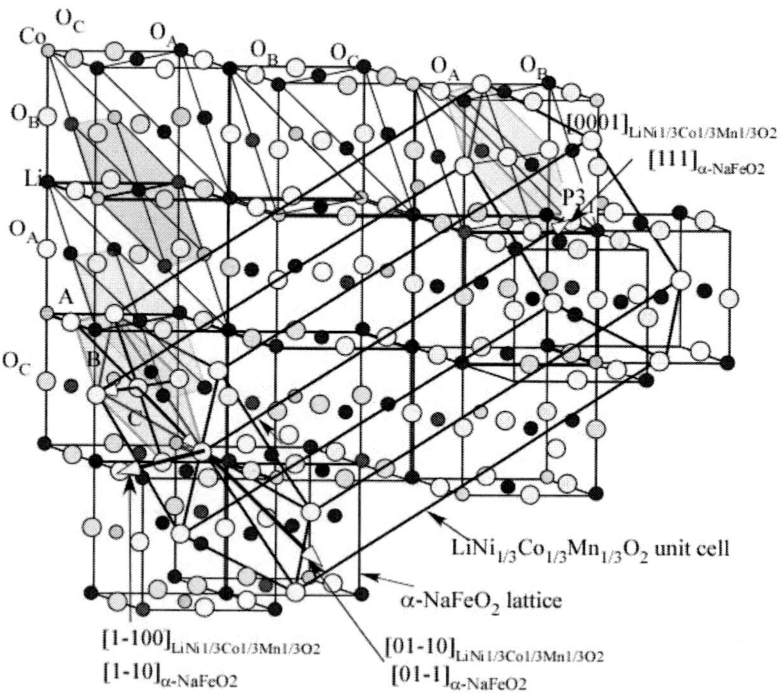

Figure 3. Relationship between the ordered $Li(Ni_{1/3}Co_{1/3}Mn_{1/3})O_2$ unit cell and $NaFeO_2$ sublattice.

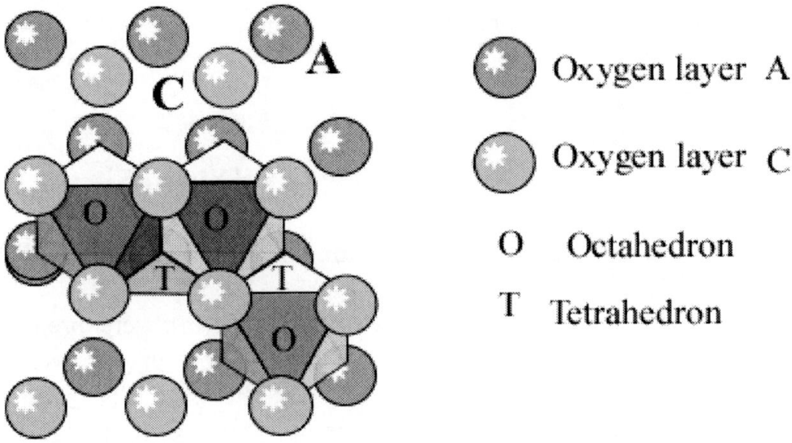

Figure 4. Successive oxygen closed packed layers form octahedron and tetrahedron located between two octahedra sharing an edge.

1.4 Analytical High Resolution Transmission Electron Microscopy (AHRTEM) is a Powerful Tool for Revealing Composition and Structure Variation of the Cathode Particles of a Cycled Lithium Battery at Atomic Scale

High resolution electron microscopy opens up the possibility of simultaneously obtain diffraction pattern on the back focus plane and images showing the projection of the atoms array of the structure of a solid on the image plane of the object lens. The electron beam size can be adjusted to as small as 5 nm to obtain the micro-diffraction and X-ray energy dispersive spectrum (EDX), which gives local structure information and chemical composition. The high resolution electron microscopy image use the coherent electron beams (or electron waves) to interact with thin solid sample and to carry the information of the distances between the atom planes, which is nearly paralleled to the electron beam direction. The diffraction spots carrying the structure information can be interfered to form the interference pattern, which may approximately reproduce the two dimension arrays of the atoms arrangement of the structure of a solid. Using object aperture in the objective lens may select different diffraction spots (or mask different diffraction spots) to interfere each other to obtain the special information of certain crystal planes of a solid. This technique is very useful for observation of Li^+ ion intercalation or de-intercalation. We use this technique to reveal the structure change of the cathode dioxide.

2. BASIC EXPERIMENT TECHNIQUES

2.1 Preparation of the Cycled Cathode Particles for AHRTEM

The observed samples of $LiMO_2$ (M=Co, Ni, Mn) were prepared as follows: MO_2 (for example $[Ni_{1/3}Co_{1/3}Mn_{1/3}]O_2$) precursors were first synthesized using co-precipitation method. $CoCl_2$, $NiCl_2$ and $MnCl_2$ salts were used as starting materials for synthesis of the MO_2 (for example $[Ni_{1/3}Co_{1/3}Mn_{1/3}]O_2$) powders. The prepared powders were mixed with excess amount of Li_2CO_3. After the mixtures were sufficiently ground, the mixtures were heated in air at 950 °C for 12 h and followed by slowly cooling to room temperature. X-ray diffraction data shows that the prepared powders have the structure of the corresponding $LiMO_2$ (M=Co,Ni,Mn).

Positive electrodes were made by coating a paste of $LiMO_2$ (for example $Li[Ni_{1/3}Co_{1/3}Mn_{1/3}]O_2$) active material, acetylene black (as a conducting additive), and polyvinylidene fluoride (PVdF) binder (93:3.5:3.5 (wt.%)) on an aluminum foil collector. Amount of the active material loaded was 5–6 mg/cm^2. The negative electrode was prepared by mixing graphite with 10 wt % PVdF binders, and the prepared paste was coated on a copper foil. The electrodes were then dried under vacuum ($5x10^{-2}$ torr) for 24 h at 120 °C. The electrolyte used was 1M LiPF6 in a (1:1:1 (wt.%)) mixture of ethylene carbonate (EC), polycarbonate (PC), and dimethyl carbonate (DMC). The cells were assembled inside an Ar-filled dry-box and were evaluated using 053048-cells (thickness: 5 mm, width: 30 mm, height: 48 mm). The discharge/charge measurements were carried out between 2.75 and 4.2 V potential ranges at a 1 C rate using an automatic battery tester at room temperature. The specific capacity vs. cycle number is shown in Figure 5, it can be seen that after 300 cycles, the discharge capacity is about 86% of initial capacity.

After 300 discharge/charge cycling, the cell was discharged to 2 V at 0.2 C and then disassembled. In order to avoid damaging the cycled particle surfaces, Pieces of the cathode obtained from the disassembled cell were ultrasonically de-agglomerated in acetone and dispersed on a holey carbon film supported by a Cu grid for TEM observation. A JEOL-2010 Electron microscope operating at 200 kV accelerating voltage and equipped with EDX was used for imaging, electron diffraction and composition micro-analysis.

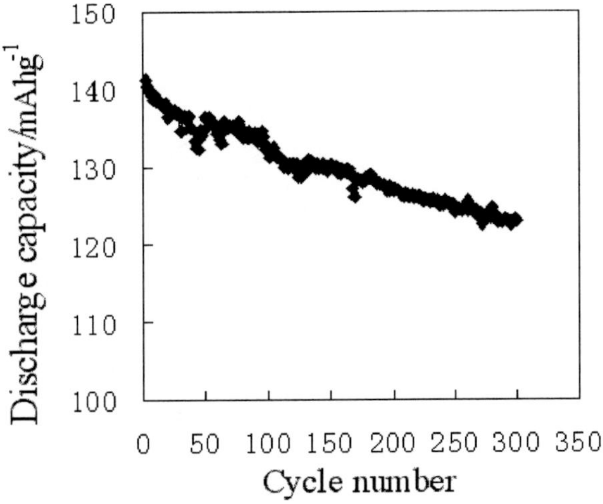

Figure 5. Discharge capacity vs. number of cycles of the cell between 2.75 and 4.2 V at 1 C rate.

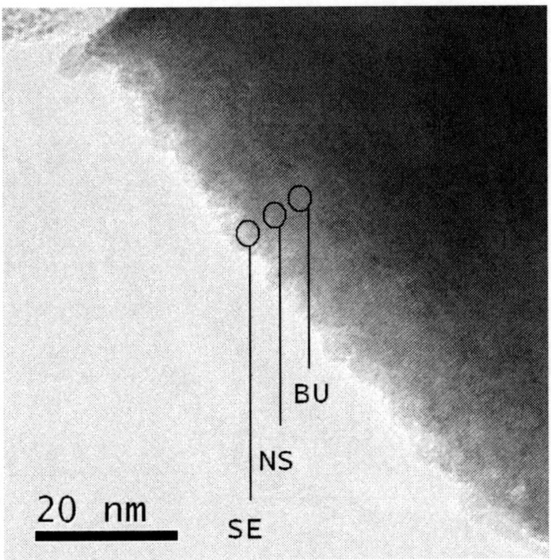

Figure 6. The different regions were micro-analyzed by EDX with 5 nm diameter electron beams.

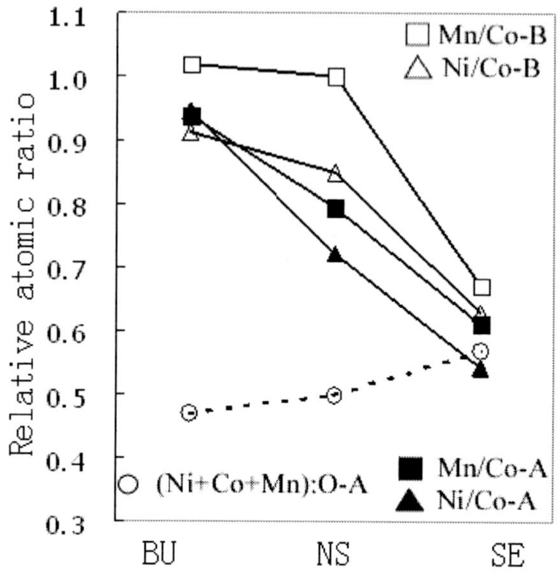

Figure 7. Relative ratios of Ni, Mn to Co in the atomic percentage decrease from BU, NS, and SE region of the cycled particles of the $Li[Ni_{1/3}Co_{1/3}Mn_{1/3}]O_2$.

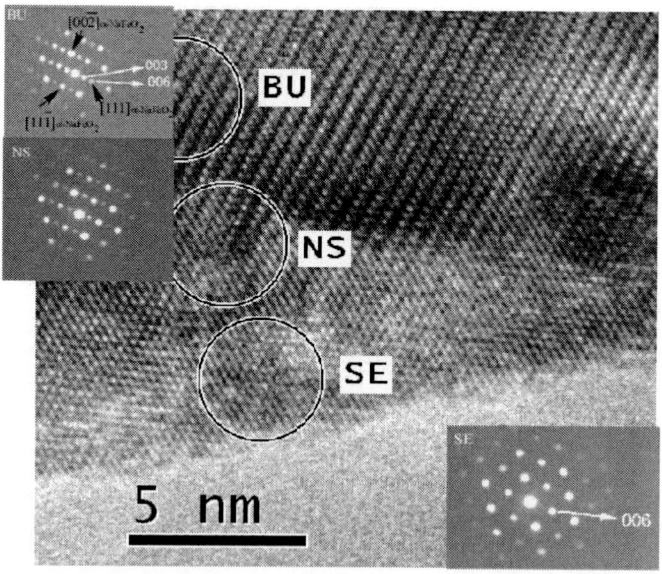

Figure 8. The micro-diffraction patterns can be obtained from the small regions.

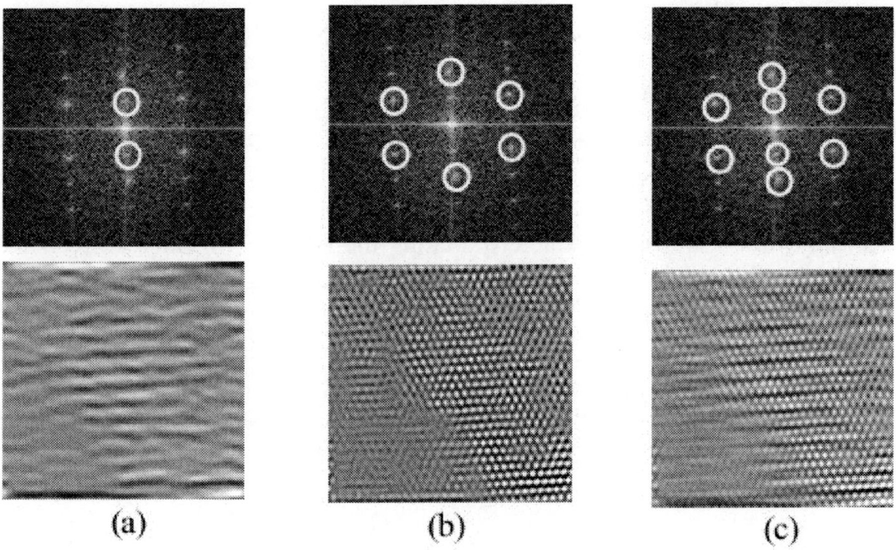

Figure 9. (a) two beams interfere (b) six stronger spots interfere (c) eight spots form the image demonstrating the structure imperfections.

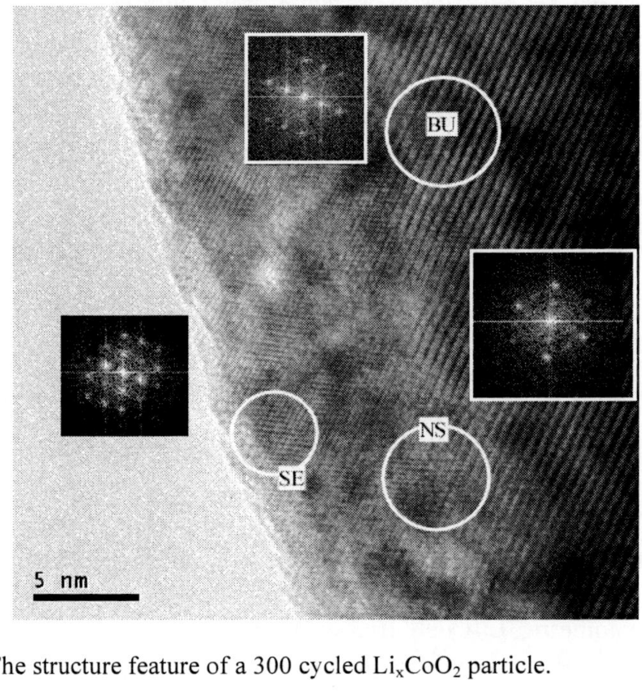

Figure 10. The structure feature of a 300 cycled Li_xCoO_2 particle.

2.2 Micro-Diffraction and Micro-Analysis of the Cycled Cathode Particles

Figure 6 shows the analyzed regions of the cycled particle. BU is the interior region. NS region is located between the edge and interior of the cycled particle. SE is placed on edge area of the cycled particle. Using an electron beam with 5 nm diameter radiating the BU, NS, SE area collects the data of EDX spectrum and the data collected from region BU were used as reference for comparison with the data collected from NS and SE area. Figure 7 gives the result of the EDX for the cycled particle of the $Li[Ni_{1/3}Co_{1/3}Mn_{1/3}]O_2$ as an example.

Figure 8 demonstrates the micro-diffraction patterns of these BU, NS, and SE small regions and the structure information of these regions may obtain.

2.3 One- and Two- Dimension Lattices Images and Analysis

As mentioned early, using different diffraction spots to form interfered image may approximately show the special atoms planes in a crystal except the objective lens of electron microscope could introduce some noise that should be keep in mind [7-10]. Figure 9 illustrates this method. The diffraction pattern shown at the top of this figure and the white circles indicate the diffraction spots to be used imaging. The bottom of this figure presents the corresponding images. In figure 9a the image reveals the one dimension lattice planes are not perfect and heavenly distorted. Figure 9b exposes the basic lattice nets of this crystal and there is some local mismatch. Figure 9c combines these two information demonstrating the unit cell and structure Imperfections.

3. Compositional and Structural Evolution of the Cathode Cycled Particles of the Lithium Batteries

3.1 $LiCoO_2$

Nonstoichiometric Li_xCoO_2 ($0.5<x<1$) has hexagonal structure, but Li^+ decrease below $x=0.5$ it will change to spinel structure, so the available Li_+

source is x=0.5. If Li content decreased more than x=05, the oxygen stacking sequence will also changed from ABC to AB or AC and that can lead the metal-insulate transition [10,11]. If Li is completely extracted from $LixCoO_2$ dioxide, It may change to CoO oxide with f.c.c structure. Figure 10 present structure change of a 300 cycled Li_xCoO_2 particle. At surface region Li is de-intercalated and some area start to form f.c.c structure. It was known [12] that $AlPO_4$ coated on the Li_xCoO_2 particle can promote to form spinel structure. Figure 11 demonstrate this factor. The surface layer of the $AlPO_4$ coated Li_xCoO_2 particle form spinel structure that may improve the Li^+ battery performance.

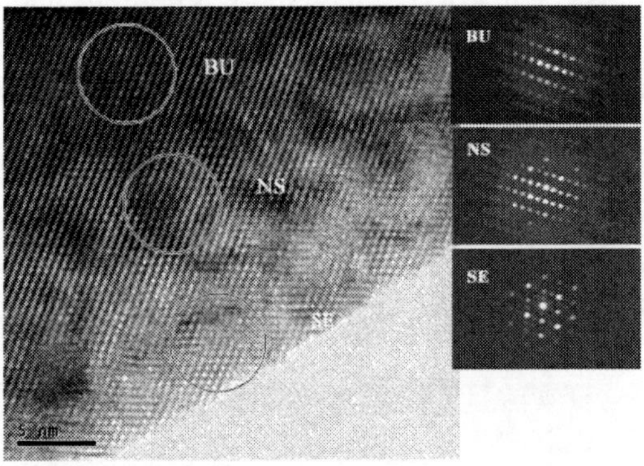

Figure 11. $AlPO_4$ coated Li_xCoO_2 particle after 300 cycling process.

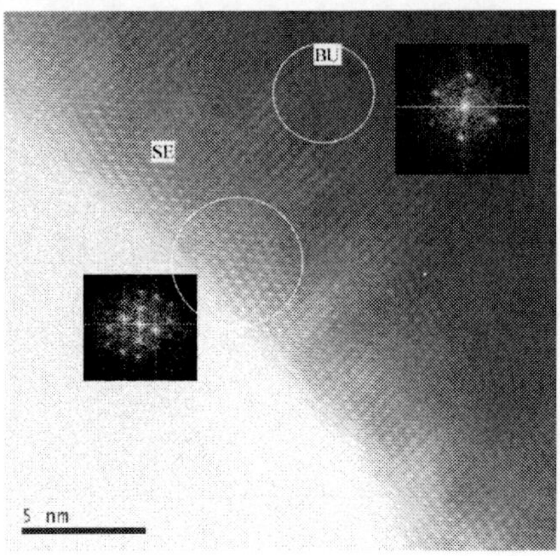

Figure 12. AlPO4 coated $LiNi_{1/3}Co_{1/3}Mn_{1/3}O_2$ cathode particle after 300 cycling process shows spinel structure on the surface region.

3.2 $LiNi_{1/3}Co_{1/3}Mn_{1/3}O_2$

The Ni and Mn in the $LiNi_{1/3}Co_{1/3}Mn_{1/3}O_2$ cathode [5,6] are easy to be lost and the surface structure change to f.c.c structure [13,14] as figure 8 shown. Using AlPO4 coated on the $LiNi_{1/3}Co_{1/3}Mn_{1/3}O_2$ cathode may also form spinel structure. Figure 12 demonstrate this result.

3.3 $LiNi_{0.8}Co_{0.2}O_2$

The $LiNi_{0.8}Co_{0.2}O_2$ cathode reduces the Co content and may increase the capacity of Li battery. But Ni^{3+} may be disproportionation into Ni^{4+} and Ni^{2+}. Ni^{2+} ions are favor in a tetrahedron site producing the spinel structure. Using $AlPO_4$ coated on the $LiNi_{0.8}Co_{0.2}O_2$ may promotes to form the spinel structure and depress the Ni^{3+} disproportionation reaction. Figure 13 show the surface structure of the $AlPO_4$ coated on the $LiNi_{0.8}Co_{0.2}O_2$. It seems that the spinel structure on the surface region may pillar up the octahedral layers resulting the improvement of the Li battery performance.

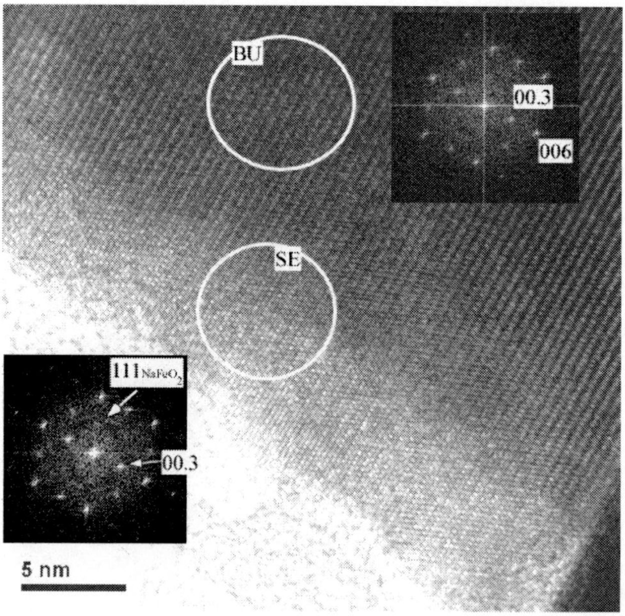

Figure 13. The $AlPO_4$ coated on the $LiNi_{0.8}Co_{0.2}O_2$ particle after 300 cycling process.

4. Discussion and Conclusion

Oxygen closed packed array is the basis of the valent variation of high-valent transition metals as Li is inserting or extracting. The d electrons of the transition metal, especially Co, Ni, Mn, could exchange their electrons with banded oxygen atoms to be reduced or oxidized by the inserted or extracted Li ion. Li migration forced by the electric field have to simultaneously accompany electron or hole navigation. Two-dimensional octahedral layer created by successive stacking (as ABC or AB or BC) of oxygen closed parked arrays is basic structure element for most cathode materials. The high valent cations are distributed in these octahedral slabs and alternately separated by the Li ion sheets. The cations in the octahedral slabs may be ordered and changed the valence state that may cause the rate of Li ion migration to vary. If the valence of a cation is changed to make this cation to be preferred in tetrahedron site created by edge shared octahedron, the structure may change to spinel, having the 2:1 ratio of octahedron and tetrahedron, or Olivine structure, having 1/2 octahedron site occupied by cations and 1/8 tetrahedron site occupied by phosphorus anions. Because the large size of the phosphorus anion can make the distance between the high valent cation octahedron layers to be large, then Olivine structure may promote the Li^+ ion migration. The $LiFePO_4$ is an example. Therefore, the cathode materials are family of the oxygen closed packed arrays. Monitoring the structure evolution of the cathode materials is the key to making a high quality Lithium battery. These targets can be comprehensively done by AHRTEM.

Acknowledgments

This work was supported by the National Natural Science Foundation of China, Grant 50372058 and the Analysis & Measurerment of Zhejiang Province Grant 2008F70050. Authors also greatly appreciate the financial support of Jinhe new material Inc.

References

[1] John, B. (2002). Goodenough “Oxide cathodes”, In: *Advances in*

Lithium ion batteries 135 edited by W. Van. Schalkwijk and B. Scrosati Kluwer Academic/Plenum Publisher.
[2] Rickert, H. (1985). *Solid State Electrochemistry, An Introduction.* Spring-Verlag, Berlin, Heidelberg, New York.
[3] Weppner, W. (1985). In "Transport-structure relations in Fast ion and Mixed Conductors" edited by F. W. Poulsen, N. Hessel Andersen, K. Clausen, S. Skaarup, O. Toft Sorensen. Riso natl Lab. Roskilde, DK, 199.
[4] Wagner 7th meeting, Intl, C. (1955). Committee on Electrochemical Thermodynamics and Kinetics Lidau Butterworths, London, 361.
[5] Ariyoshi, K. Iwakoshi, Y. Nakayama, N. Ohzuku, T. *Electrochem. J.* (2004). *Soc.*, 151, A296.
[6] Yabuuchi, N. Koyama, Y. Nakayama, N. Ohzuku, T. (2005). *J. Electrochem. Soc.*, 152, A 1434.
[7] Cowley, JM. *Diffraction Physics.* North-Holland, Amsterdam.
[8] Spence, JCH. (2003). *High resolution Electron Microscopy.* Oxford Univ. Press. Oxford and New York.
[9] Lipson, HS. (1973). *Optical Transforms*. Academic Press New York.
[10] Van der Ven, A. Aydinol , MK. Ceder, G. Créese, G. hafner, J. (1998). *Physical Review B., Vol. 58, No.6*, 2975.
[11] Gerbrand Ceder and Antón Van de Ven, (1999). *Electrochmica Acta.,* 45, 131.
[12] Youngil Lee, Min Gyu Kim, Jisuk Kim, Yoojin Kim, and Jaephil Cho, (2005). *Journal of the electrochemical society*, 152 (9), A1824.
[13] Yuewu Zeng. (2008). *Journal of Power Sources*, 183, 316-324.
[14] Yuewu Zeng, Jinghua He. (2009). *Journal of Power Sources,* 189, 519-521.

In: Lithium Batteries: Research, Technology… ISBN: 978-1-60741-722-4
Editors: Greger R. Dahlin, et al.

Chapter 7

SOFT SOLUTION PROCESSING OF NANOSCALED LITHIUM VANADIUM OXIDES AS CATHODE MATERIALS FOR RECHARGEABLE LITHIUM ION BATTERIES

Hao Wang,* HaiYan Xu and Hui Yan
The College of Materials Science and Engineering,
Beijing University of Technology, Beijing, China.

ABSTRACT

Lithium vanadium oxides have been extensively studied because of their possible application as a cathode material for rechargeable lithium batteries. Due to their low cost, they are one of the promising substitutes for the expensive $LiCoO_2$ cathode presently commercially used. Lithium vanadium oxides including γ-LiV_2O_5 and LiV_3O_8 have been prepared by soft solution methods in this study. In the first part of this work, γ-LiV_2O_5 nanorods have been prepared directly by a simple solvothermal method using ethanol as a solvent, which also serves as a reducing agent. The γ-LiV_2O_5 nanorods with diameters of 30-40 nm obtained at 160 °C shows a larger capacity of 259 mAh/g in the range of 1.5 - 4.2 V, and its capacity remained 199 mAh/g after 20 cycles. In the second part, LiV_3O_8 nanorods

* Corresponding author: E-mail: haowang@bjut.edu.cn

have been obtained by a novel hydrothermal-based two-step method. The LiV_3O_8 sample treated at 300 °C shows a poor crystallinity while a specific capacity of 302 mAh/g in the range of 1.8 - 4.0 V, and its capacity remained 278 mAh/g after 30 cycles. It indicates that the lithium vanadium oxide nanorods prepared by the above methods have potentiality to be used as cathode material in rechargeable lithium ion batteries.

INTRODUCTION

Worldwide research and development is in progress for rechargeable lithium ion batteries because of its wide application in portable electronic goods, electric vehicle systems and dispersed-type energy storage system. Much of this effort is focused on developing the cathode material for lithium ion batteries because the performance and cost of the batteries are often decided by the properties of the cathode material. In the last 20 years, vanadium oxides have been widely investigated as cathode materials in lithium ion batteries due to their low cost, low toxicity, high specific energy, and long cycle life. Lithium ion can react with V_2O_5 to form various compounds such as LiV_3O_8 [1-5], LiV_2O_5 (LiV_2O_5 has many modifications including α-, β-, ω-, γ-LiV_2O_5) [6-10], LiV_2O_4 [11], $Li_{0.6}V_{2-\delta}O_{4-\delta}$ [12] etc. Among them, γ-LiV_2O_5 and LiV_3O_8 were most intensively studied [1-10, 13-15].

The preparation method has significant influences on the electrochemical behavior of γ-LiV_2O_5 and LiV_3O_8, therefore a great deal of work has been done on their preparation to improve their electrochemical performance. In this study, soft solution methods including a solvothermal route and a modified hydrothermal-based method have been performed to prepare γ-LiV_2O_5 and LiV_3O_8 nanorods, respectively.

$Li_{1+x}V_3O_8$

Introduction

LiV_3O_8, as a species of the lithiated vanadate family, has very attractive characteristics such as high specific energy, good rate capacity, and long cycle life. The structure of LiV_3O_8 can be shown as in Figure 1. [16] It has a monoclinic structure and the space group in $P2_1/m$. The structure consists of VO_6 octahedra and VO_5 distorted trigonal bipyramids and is built up by sharing the edges and corners of the trigonal bipyramids and the octahedra to form a zigzag ribbon. Lithium can be localized in both of two kinds of vacant sites named octahedral and tetrahedral sites (Figure 1), in which the original combined lithium atoms occupy octahedral coordinated sites and the intercalated ones occupy tetrahedral sites, respectively.

It is well known that the preparation methods and post-treatments have significant influences on the electrochemical properties of LiV_3O_8. The conventional method was high temperature melting in which LiV_3O_8 was produced by reaction between Li_2CO_3 and V_2O_5 at 680 °C [17, 18]. This method has met difficulty to control the composition and homogeneity of the final products. Meanwhile, the product LiV_3O_8 had a low capacity of 180 mAh/g in the range of 1.8-4.0 V. Afterwards, many improved solution methods were proposed [2, 4, 19-21]. The solution method does not need a high reaction temperature, and the products could reach a high homogeneity and high capacity.

As one of the solution methods, hydrothermal method has been extensively used for the synthesis of inorganic compounds. Chirayil et al. had obtained a new layered lithium vanadium oxide $Li_xV_{2-\delta}O_{4-\delta}\cdot H_2O$ via hydrothermal method using tetra-methyl ammonium as template [11]. Oka et al. had prepared AV_3O_8 (A=K, Rb, and Cs) by the hydrothermal treatment of V_2O_5 and ANO_3 solutions at 250 °C. The results showed that RbV_3O_8 had the known structure of the KV_3O_8 and CsV_3O_8 compounds [22, 23]. However, attempts to make LiV_3O_8 and NaV_3O_8 hydrothermally were unsuccessful.

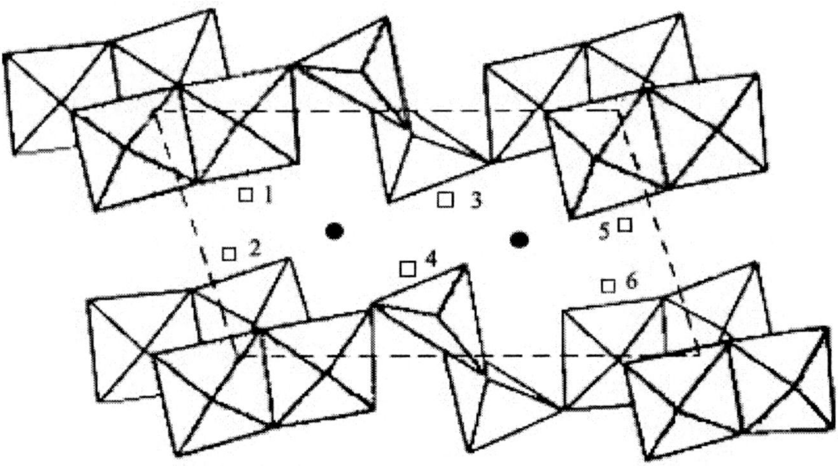

Figure 1. Structure model of LiV_3O_8. Key: (●) octahedral and (□) tetrahedral.

Some reports have also noticed the importance of crystallinity on the electrochemical properties of LiV_3O_8. It was shown that the poorly crystallized LiV_3O_8 or LiV_3O_8 modified by ultrasonic treatment or hydrothermal introduction of small amounts of inorganic compounds such as H_2O, CO_2 and NH_3 achieved an improved electrochemical performance [3, 5]. It is known that the electrochemical behavior of cathode materials strongly depends on the particle size: the bigger the particle size, the higher the cell polarization and the lower the cell capacity [8]. In the present paper, LiV_3O_8 nanorods have been prepared by a novel method, which is based on the hydrothermal reaction, the experimental procedure and results are described below.

Experimental Section

Synthesis and characterization of LiV_3O_8

The starting materials were analytically pure LiOH, V_2O_5 and $NH_3{\cdot}H_2O$ (1 mol/L). First, sotoichiometric LiOH and V_2O_5 (Li : V = 1 : 3, molar ratio) were blended in the deionized water. LiOH was dissolved completely and part of V_2O_5 was dissolved. When $NH_3{\cdot}H_2O$ was added to the above mixture, V_2O_5 was dissolved completely into solution. The pH of the solution was 9. The resultant dark green solution was then transferred to a 50 ml Teflon lined autoclave. The autoclave was sealed and heated at 160 °C under auto-generated pressure for 12 h. After hydrothermal treatment, a colorless clear solution whose pH changed to 7 was obtained. This solution was dried in air at 100 °C to evaporate the water till an orange gel was prepared. The gel was then heat-treated at different temperatures in the range of 300-600 °C for 12 h.

The gel was characterized by thermo-gravimetric analysis (TGA) using a Model STA 449C (Germany, NETZSCH-Gerätebau GmbH Thermal Analysis). The structure of the heat-treated products was examined by X-ray diffractometry (XRD, Japan Rigaku D/Max-3C) using CuKα radiation (λ = 1.5405Å). The morphology was investigated by transmission electron microscopy (TEM, Model Hitachi H-700H, 200 KV). Fourier transform infrared (FTIR) absorption spectra were obtained by using a Nicolet Magna-IR 560 spectroscopy.

Electrochemical measurements

Electrochemical characterization of the products was performed in cells with metallic lithium as the negative electrode and a liquid organic electrolyte

[$LiPF_6$ in a volume ratio of 50:50 mixture of ethylene carbonate (EC) and diethyl carbonate (DEC), absorbed in porous polypropylene separators, Celgard 2400]. The cathode was a mixture of the active material, acetylene black and poly (tetrafluoroethylene) (PTFE) in a weight ratio of 80:10:10. The cells were assembled in an argon-filled dry box. Charge-discharge tests were carried out at a constant current density of 0.3 mA/cm^2 in a range of 1.8-4.0 V. All the tests were performed at room temperature.

Results and Discussion

The TGA result of the precursor gel which was derived by hydrothermal reaction and subsequent evaporation of water is illustrated in Figure 2. It can be seen that in the range of 140 ~ 320 °C, the weight loss is about 15 %. Above 320 °C, the weight remains stable up to 600 °C. The weight loss process can be divided into two stages. The first weight loss begins at 140 °C and ends at 230 °C with a weight loss of 13%, which mainly resulted from the evaporation of NH_3 and water. The second weight loss occurred in the range of 250 ~ 320 °C with a weight loss of 2 %, which was caused by the de-intercalation of some strongly-bound water.

TGA result

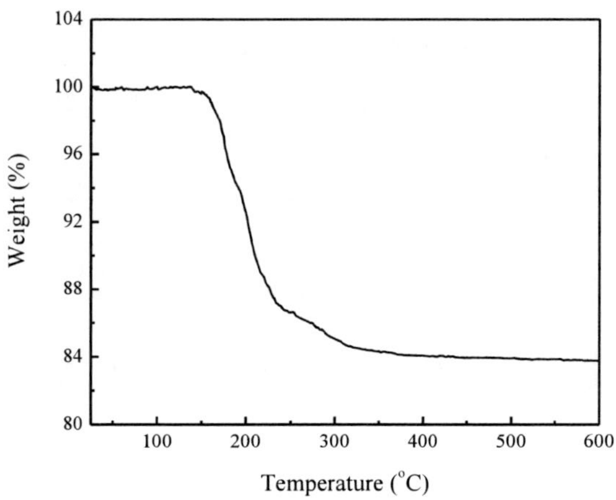

Figure 2. TGA curve of the precursor gel.

The XRD and the Structure of LiV_3O_8

Figure 3 shows the XRD patterns of the products heated at 300, 350, 400 and 600 °C. These XRD patterns reflect the structural variation during the treatment process. With an increase in heat temperature the intensity of peaks becomes stronger and the full width of half maximum intensity (FWHM) decreases, which indicates that the crystallinity becomes higher.

In addition, it is noted that there is an obvious difference among the four XRD patterns. The relative intensity of (100) peaks at around 13.86° becomes stronger with increasing heat temperature. In the 600 °C diffraction pattern, the intensity of (100) peak is stronger than any other peaks. While in the 300 °C diffraction pattern, the relative intensity of (100) peak has much decreased. It suggests that the particle shape of LiV_3O_8 depends on the heat temperature, and higher heat temperature favors the preferential ordering of crystallites. It is known that the intercalation process of Li^+ ion between the layers of the cathode is a diffusion process. Therefore the preferential ordering which would lead to a long path for Li^+ ion is not advantageous to intercalation [2].

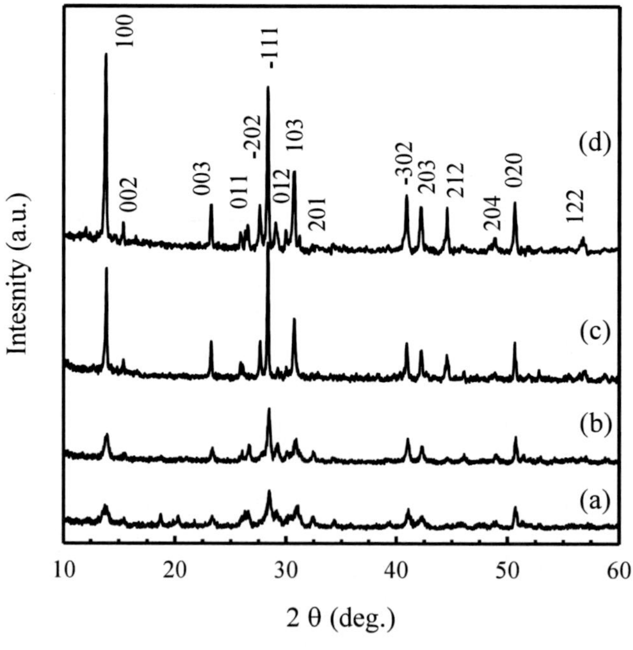

Figure 3. The XRD patterns of the samples heat treated at (a) 300, (b) 350, (c) 400, (d) 600 °C.

(a)

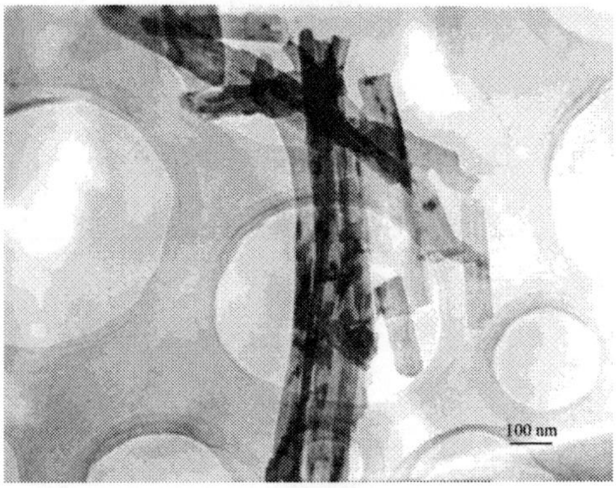

(b)

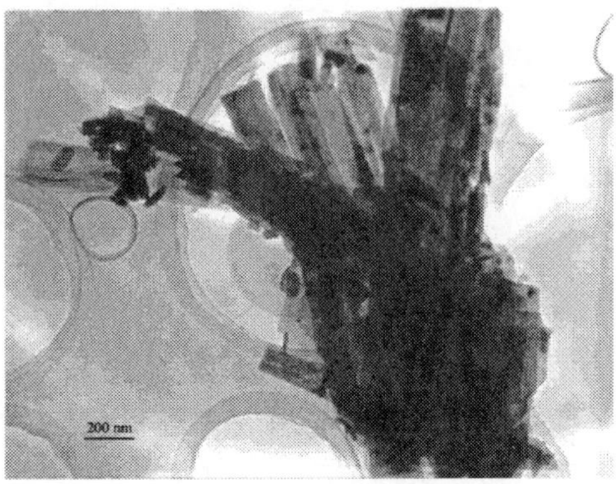

(c)

Figure 4. TEM micrographs of the samples heat treated at (a) 300, (b) 350, (c) 400 °C.

The morphology of the as-synthesized LiV_3O_8

The TEM micrographs of the synthesized LiV_3O_8, at different heat temperatures, are demonstrated in Figure 4. It can be seen that the heat treatment has caused a change in the LiV_3O_8 crystallinity and morphology. The sample treated at 300 °C (Figure 4a) consists of an agglomeration of small rods. The shapes of the rods are not well recognized, indicating they are relatively poorly crystallized. The diameters of the rods are about 40 nm and the lengths are mostly less than 600 nm. Figure 4b is the micrograph of the sample treated at 350 °C. Comparing to the 300 °C sample, the 350 °C sample is well rod-shaped. The diameters of the rods are about 70 nm while the lengths are diverse in the 0.5-2 μm range. The 400 °C sample which is shown in Figure 4c consists of wider (>150 nm) and longer rods. The sizes of the particles treated at 600 °C are much larger than that of the 400 °C sample, the micrograph of which is not shown here. These results are in agreement with the XRD data. With the increase in temperature, the particles of the products become larger and more crystallized.

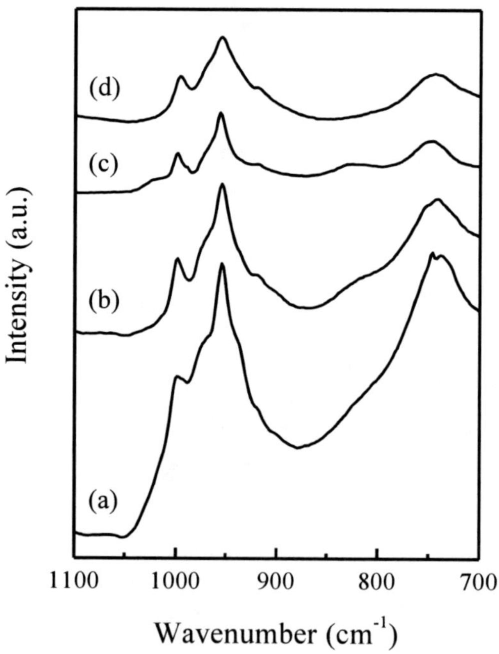

Figure 5. The FTIR spectra of the samples heat treated at (a) 300, (b) 350, (c) 400, (d) 600 °C.

FTIR of the as-synthesized LiV_3O_8

FTIR spectra of LiV_3O_8 obtained at different heat temperatures are shown in Figure 5. For the four curves, it can be seen that the FTIR absorption in the spectral region of 700 ~ 1100 cm^{-1} is dominated by bands at 995, 954 and 744 cm^{-1}, respectively. This data is compared with LiV_3O_8 synthesized from solid state reactions, where major FTIR bands corresponding to V = O and V-O-V vibrations are located at 996, 954 and 746 cm^{-1}, respectively [24]. The sample treated at 300 °C gives the same FTIR responses as that of crystalline LiV_3O_8, suggesting that it is almost iso-structural with the latter.

Electrochemical properties of the as-synthesized LiV_3O_8

The charge-discharge curves at the second cycle of cells from products heat treated at 300, 350, 400, 600 °C, respectively, are illustrated in Figure 6. It is found that the specific capacities of the samples decrease with increasing the heat temperature. The discharge capacity for the 300 °C sample is 302 mAh/g in the range of 1.8 - 4.0 V, which is much higher than that of 600 °C sample (190 mAh/g). Notably, the value of 302 mAh/g is considerably higher than the capacities of 220 ~ 274 mAh/g of LiV_3O_8 synthesized by other solution methods [2, 4, 19-21], and exceeds the capacity of about 280 mAh/g of hydrothermally and ultrasonically treated LiV_3O_8 [3, 5]. From the XRD data we have known that the sample treated at higher temperature has higher crystallinity. The above results also suggest that the well-crystallized sample does have poor specific capacity, which is in agreement with the previous studies [2-5, 19-21].

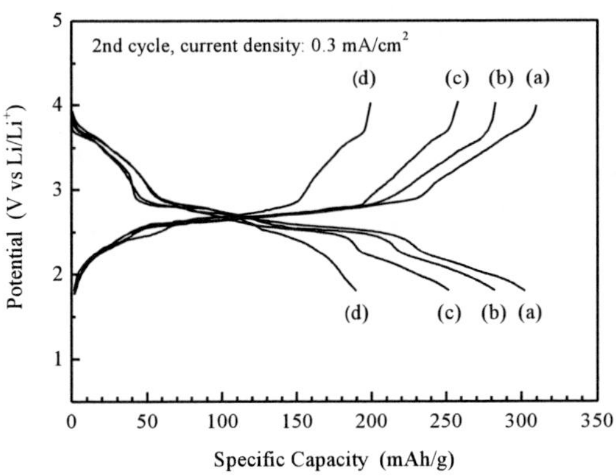

Figure 6. Charge-discharge curves of the samples heat treated at (a) 300, (b) 350, (c) 400, (d) 600 °C. Current density: 0.3 mA/cm^2.

In Figure 7, the discharge capacities are shown as a function of cycle number. It can be seen that the samples synthesized via this novel method show better capacity retention. The capacity of the 300 °C samples is 278 mAh/g after 30 cycles. Meanwhile, the 300 °C sample shows better cycling behavior compared with the samples treated at higher temperature.

CONCLUSION

In this study, a novel method, which is based on the hydrothermal reaction has been performed to synthesize LiV_3O_8 nanorods. This hydrothermal reaction did not directly yield a solid product. A gel, which was used as the precursor for the post heat treatment was obtained after evaporation of the hydrothermal treated solution. The gel was homogeneous and ultrafine, which should be the reason why the LiV_3O_8 nanorods could be obtained in this study. Heat treatment at different temperatures influenced the particle size and crystallinity of the products, which consequently affected their electrochemical performance. The sample treated at 600 °C shows a good crystallinity and a low discharge capacity. In contrast, the sample treated at 300 °C shows a poorer crystallinity while a better capacity of 302 mAh/g in the range of 1.8 - 4.0 V, and its capacity remained 278 mAh/g after 30 cycles.

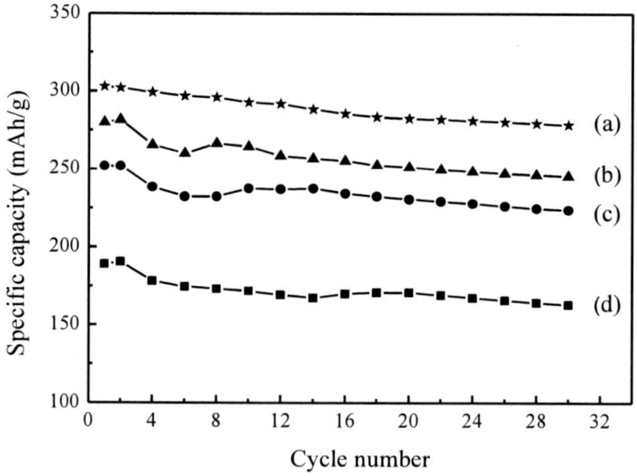

Figure 7. The cycle performance of the cells with LiV_3O_8 heat treated at (a) 300, (b) 350, (c) 400, (d) 600 °C as cathode active material. Current density: 0.3 mA/cm^2. Voltage window: 1.8-4.0 V.

γ-LiV_2O_5

Introduction

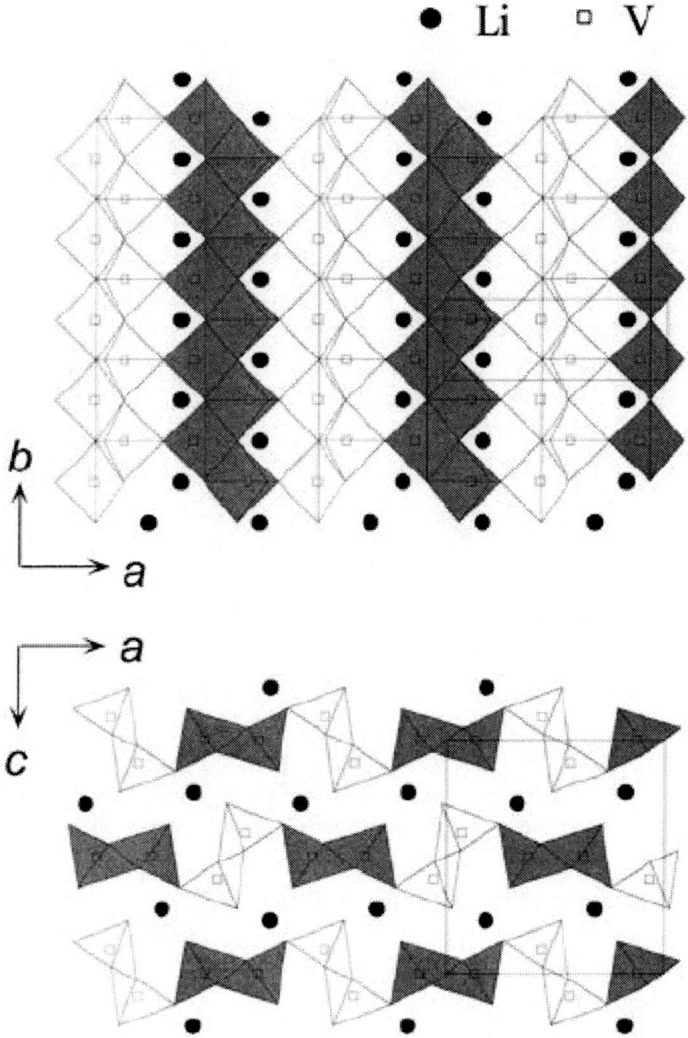

Figure 8. Schematic crystal structure of γ-LiV_2O_5 projected onto the orthorhombic *a-b* plane and *a-c* plane. The filled circles represent Li^+ ions. The squares represent V atoms. The white and shaded square pyramids show two kinds of VO_5 pyramids (two crystallographic vanadium sites).

γ-LiV_2O_5, as one of the promising cathode materials for rechargeable lithium batteries, has attracted tremendous research interest in recent years [6, 9, 10]. Orthorhombic γ-LiV_2O_5 has a layer structure with lithium ions between the layers. In this structure, there are two crystallographic vanadium sites that form two kinds of zigzag chains, the shaded and the white zigzag chains, as shown in Figure 8 [25]. The valence states of the vanadium ions were inferred from the results of structural analysis to be V^{4+} for the shaded zigzag chains and V^{5+} for the white zigzag chains [25]. Within the layers, $V^{4+}O_5$ (shaded) zigzag chains are linked to $V^{5+}O_5$ (white) zigzag chains by corner sharing, as shown in Figure 8.

In particular, a lot of work has been devoted to the synthesis of γ-LiV_2O_5, because it is very difficult to synthesize this compound, in which there are two valence states (V^{5+} and V^{4+}) of vanadium co-existing. So far, solid-state reaction routes[2] and a procedure developed by Murphy et al. [7] were the two commonly used preparation method for the synthesis of γ-LiV_2O_5. By solid-state reaction routes, the mixture of $LiVO_3$, V_2O_5, and VO_2 or V_2O_3 was heated at above 600 °C under vacuum or nitrogen protected atmosphere in order to control the valence balance of V^{4+} and V^{5+}. This calcinating method is usually faced with difficulty to control the homogeneity and particle size of the final products. It has been shown that the electrochemical behavior of cathode materials strongly depends on the particle size: the bigger the particle size, the higher the cell polarization and the lower the cell capacity [8]. The synthetic method developed by Murphy et al. includes two steps: firstly, δ-LiV_2O_5 was synthesized through the reduction of V_2O_5 with appropriate amount of lithium iodide in acetonitrile (CH_3CN) at ambient temperature; secondly, the as-prepared δ-$Li_xV_2O_5$ was heated at 350 °C under argon to obtain γ-LiV_2O_5. Such route has demonstrated the feasibility of preparing γ-LiV_2O_5 as a cathode material for rechargeable lithium ion batteries. However, the high cost and sensitivity to moisture of LiI and CH_3CN make it not viable as a low cost and low toxicity preparation method for cathode materials.

We have proposed a simple and mild solvothermal method for the synthesis of γ-LiV_2O_5. In this process, γ-LiV_2O_5 nanorods were synthesized directly from the solvothermal reaction of V_2O_5, LiOH and ethanol at 160-200 °C in an autoclave. Ethanol was employed as a solvent as well as a reducing agent. Compared with the two commonly used preparation methods for γ-LiV_2O_5, the newly developed solvothermal method is cheaper and milder, for example, vaccum or argon/nitrogen protected atmosphere or post annealing is not necessary for this simple one-step process. Therefore, the solvothermal process seems to offer a potentially low-temperature, low-cost and environmentally friendly way of producing single-phase, uniform-particle size and fine-grained γ-LiV_2O_5 for rechargeable lithium batteries. In this paper, the effect of experimental parameters including temperature and time on the properties of γ-LiV_2O_5 was also investigated. The as-obtained products were further characterized by X-ray diffraction (XRD), Fourier transform infrared (FTIR) spectroscopy, X-ray photoelectron spectrscopy (XPS), and transmission electron microscopy (TEM). The electrochemical performance of γ-LiV_2O_5 nanorods with different size was studied.

Experimental Section

Synthesis and characterization of γ-LiV_2O_5

In a 50 ml Teflon vessel, 0.02 mol of analytically pure LiOH and V_2O_5 were mixed in 40 ml of ethanol. Be well stirred, the Teflon vessel containing the mixture was put into a stainless steel autoclave. After the autoclave was kept at 160, 180, 200 °C under autogeneous pressure for 12-48 h, it was allowed to cool to room temperature naturally. The as-formed solid precipitate was filtered, washed with ethanol, and dried at 100 °C for 2 h.

For all products, XRD (Japan Rigaku D/Max-3C, Cu Kα radiation) was utilized to identify the produced phase. Fourier transform infrared (FTIR) absorption spectra were obtained by using a Nicolet Magna-IR 560 spectroscopy. XPS measurements were performed on a PHI-5300/ESCA system with Mg Kα radiation as the exciting source, where the binding energies were calibrated by referencing the C 1s peak to reduce the sample charge effect. The morphology and particle sizes of the as-obtained γ-LiV_2O_5 were determined by TEM (Model Hitachi H-700 H, 200 KV).

Electrochemical measurements

Electrochemical tests were performed in cells with metallic lithium as the negative electrode. The electrolyte was $LiPF_6$ in the mixture of ethylene carbonate (EC) and diethyl carbonate (DEC) (50:50), Celgard 2400 as separators. The positive electrode composites were made by mixing the active material, acetylene black and poly (tetrafluoroethylene) (PTFE) (80:10:10). All cells used in this study were assembled in an argon-filled dry box. Charge-discharge tests were carried out at a constant current density of 0.3 mA/cm^2 in a range of 1.5 - 4.2 V. All tests were performed at room temperature.

Results and Discussion

The XRD results and the kinetic processing of the formation of γ-LiV$_2$O$_5$

Figure 9 shows the XRD patterns of the black products solvothermally synthesized at 160, 180, 200 °C for 24 h. As can be seen, all the obtained products have almost the same XRD patterns; which can be assigned to an orthorhombic structure γ-LiV_2O_5 according to JCPDS cards No. 018-756. It is suggested the formation of γ-LiV_2O_5 without any trace of impurities such as V_2O_5, Li_3VO_4 and $LiVO_2$ by this solvothermal route.

Figure 10 shows the XRD patterns of the products solvothermally synthesized at 160 °C for 12-48 h. It is obvious that the variation of reaction time causes a change in the XRD patterns of the resulting products. For 12 h, the obtained product mainly consists of V_2O_5, in which the average valence of vanadium is +5; For 18-24 h, the obtained product is pure γ-LiV_2O_5, in which the average valence of vanadium is + 4.5. When the reaction time is further prolonged to 48 h, the obtained product is chiefly $Li_xV_{2-\delta}O_{4-\delta}\cdot H_2O$ [11] with a minor amount of γ-LiV_2O_5, in which the average valence of vanadium is lower than +4. The above results indicate that when the reaction temperature is fixed at 160 °C, the reaction time plays an important role in determining the phases of products. We have also investigated the effect of the reaction time on the phase(s) of the resulting products at the temperatures of 180 and 200 °C, and obtained the similar results. It is suggested that, in this study, the formation of γ-LiV_2O_5 can be attributed to a process that is mainly controlled by kinetics.

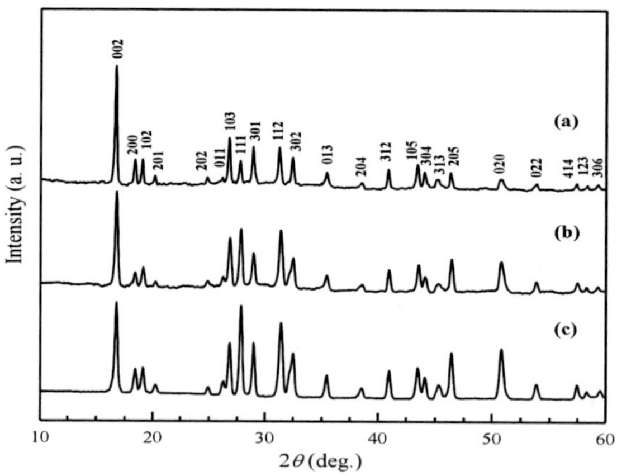

Figure 9. The XRD patterns of the samples synthesized at (a) 160, (b) 180, (c) 200 oC for 24 h.

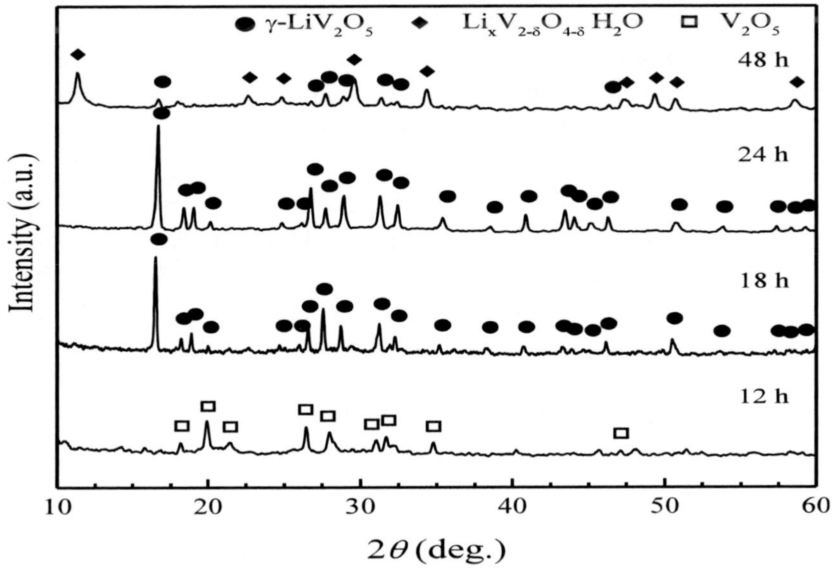

Figure 10. The XRD patterns of the samples synthesized at 160 °C for (a) 12, (b) 18, (c) 24, (d) 48 h.

In our experiment, the synthesis of γ-LiV_2O_5 is carried out using a great excess of ethanol as a solvent through a solvothermal process. To the ethanol, a portion of it is oxidized (by V_2O_5) to aldehyde, which is experimentally confirmed through chemical analysis of the final solution mixture. From the above results, it can be seen that the ethanol acts not only as a solvent but also as a reducing agent in the synthesis of γ-LiV_2O_5 powders. The possible formation mechanism of γ-LiV_2O_5 through the solvothermal reaction of LiOH and V_2O_5 in ethanol medium is a reductive recombination pathway. Thus, the formation mechanism of γ-LiV_2O_5 through the reaction of LiOH, V_2O_5 and excess ethanol under the solvothermal condition can be expressed as the following:

$$2\ LiOH + 2\ V_2O_5 + CH_3CH_2OH \rightarrow 2\ \gamma\text{-}LiV_2O_5 + CH_3CHO + 2\ H_2O$$

If the reaction time is further prolonged, the obtained γ-LiV_2O_5 may be reduced to $Li_xV_{2-\delta}O_{4-\delta}\cdot H_2O$, and the reaction equation can be expressed as the following:

$$\gamma\text{-}LiV_2O_5 + CH_3CH_2OH \rightarrow Li_xV_{2-\delta}O_{4-\delta}\cdot H_2O + CH_3CHO$$

FTIR of the as-synthesized γ-LiV_2O_5

Figure 11 shows the FTIR spectra of the as-synthesized γ-LiV_2O_5 at 160 °C for 24 h and the starting material V_2O_5. It can be seen, the FTIR absorption bands of γ-LiV_2O_5 are different from those of V_2O_5 in the spectral region of 400 ~ 1200 cm^{-1}. The FTIR spectrum of the as-synthesized γ-LiV_2O_5 (Figure 11 (a)) displays four major absorption bands at 1006, 952, 592, and 565 cm^{-1}, which are in agreement with the literature values. The absorption bands at 1006 and 952 cm^{-1} are attributed to V=O stretching vibrations, and the absorption bands at 592 and 565 cm^{-1} are attributed to V-O-V bending vibrations [26]. The FTIR spectrum of the starting material V_2O_5 (Figure 11 (b)) exhibits four major absorption bonds at 1025, 824, 600 and 473 cm^{-1} which have been associated with the V=O and V-O-V vibrations [26].

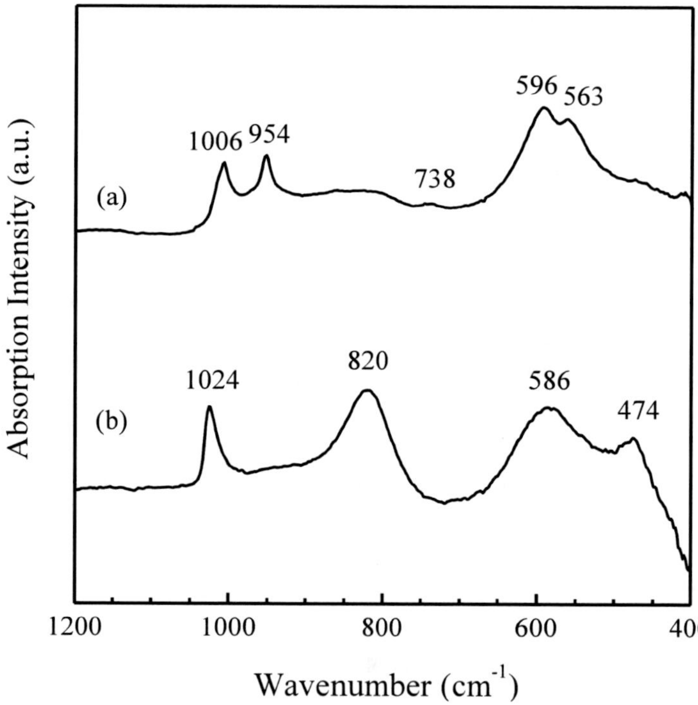

Figure 11. The FTIR spectra of (a) γ-LiV_2O_5 synthesized at 160 °C for 24 h, and (b) the starting material V_2O_5.

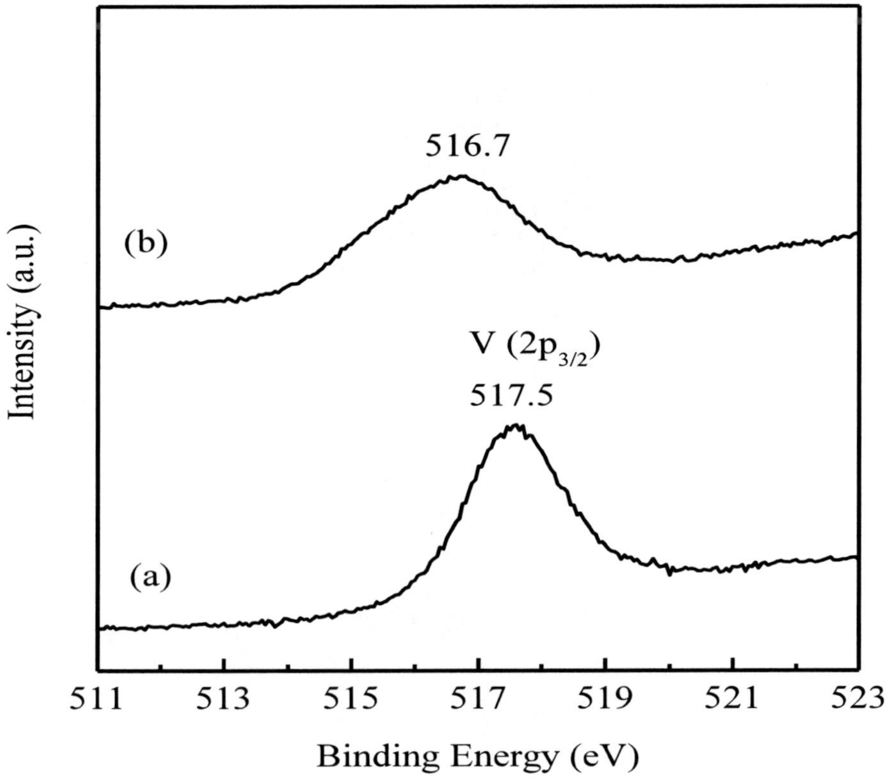

Figure 12. The XPS spectra of (a) γ-LiV_2O_5 synthesized at 160 °C for 24 h, and (b) the starting material V_2O_5.

XPS of the as-synthesized γ-LiV_2O_5

Figure 12 shows the XPS spectra of the starting material V_2O_5 and the as-synthesized γ-LiV_2O_5. It is obvious that the binding energy of V 2p for the as-synthesized γ-LiV_2O_5 is lower than that for the starting material V_2O_5. The binding energy of V $2p_{3/2}$ for the starting material V_2O_5 is 517.5eV, which agrees with the literature value of V_2O_5 [27]. The binding energy of V $2p_{3/2}$ for the as-synthesized γ-LiV_2O_5 is 516.7 eV, which is between those of V $2p_{3/2}$ for V_2O_5 and VO_2 (516.3 eV [27]). The average valence of vanadium in V_2O_5, γ-LiV_2O_5 and VO_2 is +5, +4.5 and +4, respectively.

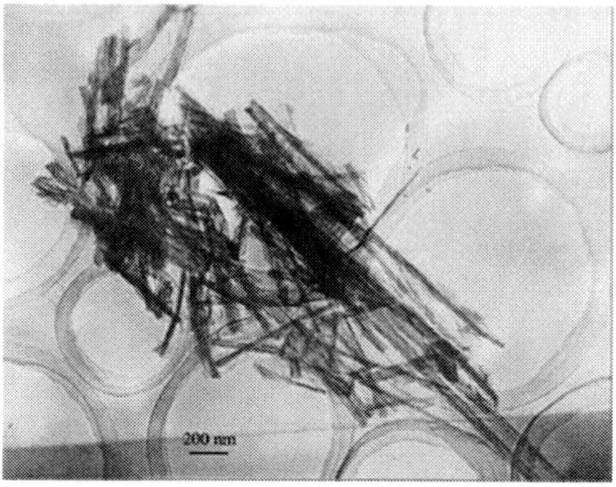

(a)

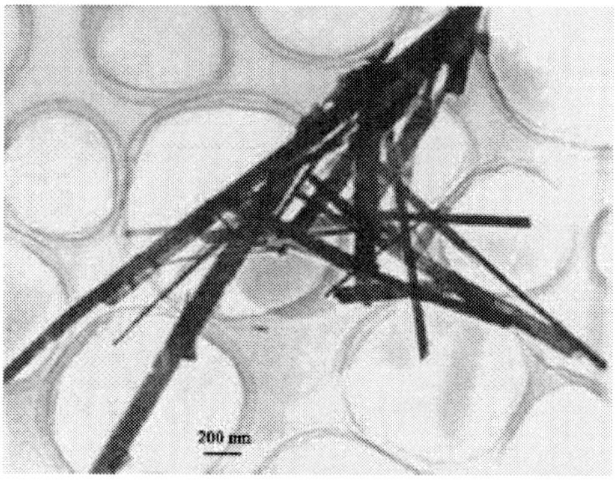

(b)

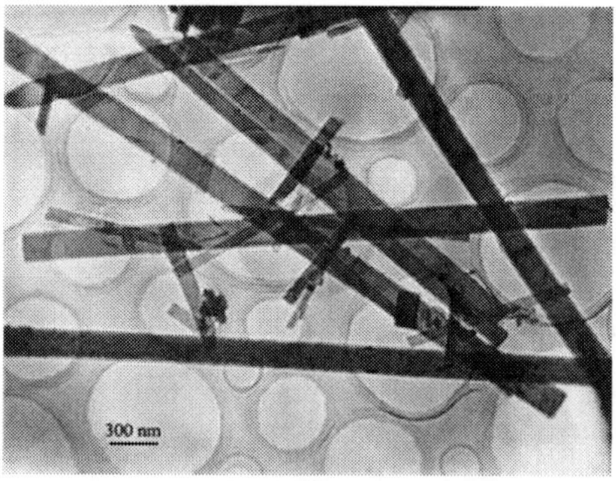

(c)

Figure 13. TEM micrographs of the γ-LiV_2O_5 synthesized at (a) 160, (b) 180, (c) 200 ^oC for 24 h.

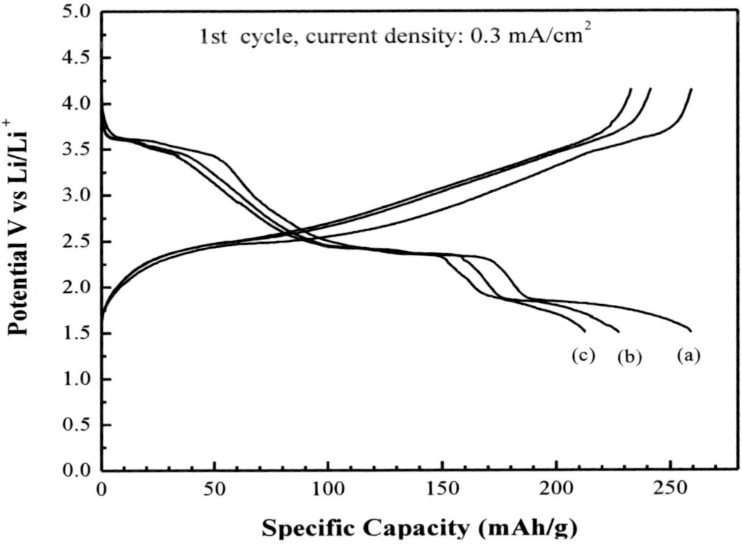

Figure 14. Charge-discharge curves of the cells with γ-LiV_2O_5 synthesized at (a) 160, (b) 180, (c) 200 °C for 24 h as cathode active material. Current density: 0.3 mA/cm^2.

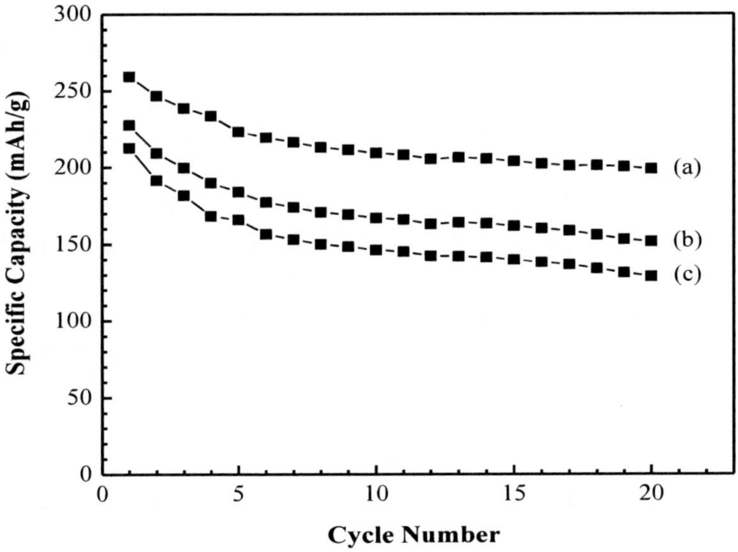

Figure 15. The cycle performance of the cells with γ-LiV_2O_5 synthesized at (a) 160, (b) 180, (c) 200 °C for 24 h as cathode active material. Current density: 0.3 mA/cm^2. Voltage window: 1.5-4.2 V.

The morphologies of the as-synthesized γ-LiV$_2$O$_5$

Figure 13 shows the TEM micrographs of the as-synthesized γ-LiV_2O_5 at 160,180, 200 °C for 24 h. As can be seen, the solvothermal reaction temperature has an effect on the particles size of the as-synthesized γ-LiV_2O_5. With the increase in solvothermal reaction temperature, the as-synthesized γ-LiV_2O_5 particles become larger and even their size distribution becomes wider. The as-synthesized γ-LiV_2O_5 at 160 °C (Figure 13a) comprises nanorods with diameters of 30-40 nm and lengths of 0.4-2 μm; The as-synthesized γ-LiV_2O_5 at 180 °C (Figure 13b) consists of wider and longer rods with diameters of 50-120 nm and lengths of 0.7-3.5 μm; The as-synthesized γ-LiV_2O_5 at 200 °C for 24 h (Figure 13c) is composed of rods with diameters of 70-160 nm and lengths of 0.7-4 μm.

Electrochemical properties of the as-synthesized γ-LiV$_2$O$_5$

Figure 14 shows the charge-discharge curves of the cell with the solvothermally-synthesized γ-LiV_2O_5 at 160, 180 and 200 °C for 24 h during the first cycle. It can be seen that the discharge capacity for the product synthesized at 160°C is 259 mAh/g in the range of 1.5 - 4.2 V, which is much higher than that of the product synthesized at 200 °C (218 mAh/g) and 180 °C (228 mAh/g). From the TEM results, we know that the solvothermal-synthesized γ-LiV_2O_5 at 200 °C has a larger size than that of the product obtained at 160 °C. Thus, it is reasonable to infer that the product with smaller particle size has a higher specific capacity, which is also consistent with the previous report [5].

Figure 15 shows the cycle performance of the cell with the as-synthesized γ-LiV_2O_5 at 160, 180 and 200 °C for 24 h. It can be seen that the solvothermally synthesized γ-LiV_2O_5 shows better capacity retention. In the first 6 cycles, the specific capacities of three cells reduce quickly, and then maintain stable in the subsequent cycles. The specific capacity for the cell with γ-LiV_2O_5 synthesized at 160 °C is 199 mAh/g after 20 cycles. The specific capacity for the cell with γ-LiV_2O_5 synthesized at 180 °C and 200 °C is 152 mAh/g and 129 mAh/g after 20 cycles, respectively. The above results indicate that the γ-LiV_2O_5 nanorods synthesized by this solvothermal method have potentiality to be used as a cathode material in rechargeable lithium cells.

CONCLUSION

In summary, LiV_3O_8 nanorods have been obtained by a novel two-step method which was based on the hydrothermal reaction. The as-prepared LiV_3O_8 nanorods showed a good charge-discharge and cycle performance. Heat treatment at different temperatures influenced the particle size and crystallinity of the products, which consequently affected their electrochemical performance. The sample treated at 300 °C shows a poorer crystallinity while a better capacity of 302 mAh/g in the range of 1.8 - 4.0 V, and its capacity remained 278 mAh/g after 30 cycles. Meanwhile, A simple and low-cost solvothermal method has been developed to synthesize γ-LiV_2O_5 nanorods by using ethanol as a solvent as well as a reducing agent. Preliminary electrochemical tests indicated that the γ-LiV_2O_5 obtained at 160 °C has an initial specific capacity of 259 mAh/g in the range of 1.5–4.2 V. It indicates that the lithium vanadium oxide nanorods prepared by the above methods have potentiality to be used as cathode material in rechargeable lithium batteries.

REFERENCE

[1] Panero, S; Pasquali, M; Pistoia, G. *J. Electrochem. Soc.*, 1983, 130, 1225.
[2] Pistoia, G; Pasquali, M; Wang, G; Li, LJ. *J. Electrochem. Soc.*, 1990, 137, 2365.
[3] Manev, V; Momchilov, A; Nassalevska, A; Pistoia, G; Pasquali, M. *J. Power Sources.*, 1995, 54, 501.
[4] West, K; Zachau-Christiansen, B; Skaarup, S; Saidi, Y; Barker, J; Olsen, II; Pynenburg, R; Koksbang, R. *J. Electrochem. Soc.*, 1996, 143, 820.
[5] Kumagai, N; Yu, A. *J. Electrochem. Soc.*, 1997, 144, 830.
[6] Delmas, H; Cognac-Auradou, JM; Cocciantelli, M; Menetrier, JP. Doumerc, *Solid State Ionics.*, 1994, 69, 257.
[7] Murphy, W; Christian, PA; Disalvo, FJ; Waszczak, JV. *Solid State Ionics.*, 1979, 18, 2800.
[8] Cocciantelli, JM; Menetrier, M; Delmas, C; Doumerc, JP; Pouchard, M; Broussely, M; Labat, J. *Solid State Ionics.*, 1995, 78, 143.
[9] Rozier, P; Savariault, JM; Galy, J. *Solid State Ionics.*, 1997, 98, 133.
[10] Dai, J; Li, SFY; Gao, Z; Siow, KS. *Chem. Mater.*, 1999, 11, 3086.
[11] Chirayil, T; Zavalij, P; Whittingham, MS. *Solid State Ionics.*, 1996, 84, 163.

[12] Chirayil, T; Zavalij, P; Whittingham, MS. *J. Electrochem. Soc.*, 1996, 143, L193.
[13] Xu, HY; Wang, H; Song, ZQ; Wang, YW; Zhang, YC; Yan, H. *Chem. Lett.*, 2003, 32, 444.
[14] Xu, HY; Wang, H; Song, ZQ; Wang, YW; Yan, H; Yoshimura, M. *Electrochimica Acta.*, 2004, 49, 349.
[15] Wang, YW; Xu, HY; Wang, H; Zhang, YC; Song, ZQ; Yan, H; Wan, CR. *Solid State Ionics.*, 2004, 167, 419.
[16] Wadsley, AD. *Acta Cryst.*, 1957, 10, 261.
[17] Pistoia, S; Panero, M; Tocci, R; Moshtev, V; Manev, J. *Solid State Ionics.*, 1984, 13, 12.
[18] Pistoia, M; Pasquali, M; Tocci, V; Manev, R; Moshtev, J. *Power Sources.*, 1985, 15, 13.
[19] Yu, A; Kumagai, N; Liu, Z; Lee, JY. *J. Power Sources.*, 1998, 74, 117.
[20] Liu, Q; Zeng, CL; Yang, K. *Electrochimica Acta.*, 2002, 47, 3239.
[21] Kawakita, T; Kato, Y; Katayama, T; Miura, T; Kishi, J. *Power Sources.*, 1999, 81-82, 448.
[22] Oka, Y; Yao, T; Yamamoto, N. *Mater. Res. Bull.*, 1997, 32, 1201.
[23] Evans, HT; Black, S. *Inorg. Chem.*, 1966, 10, 1808.
[24] Zhang, X; Frech, R. *Electrochimica Acta.*, 1998, 43, 861.
[25] Anderson, DN; Willett, RD. *Acta Crystallogr. Sect. B.*, 1971, 27, 1476.
[26] Zhang, XL; Frech, R. *Electrochimeca Acta.*, 1997, 42, 475.
[27] Wagner, CD. *Handbook of X-ray Photoelectron Spectroscopy*, Minnesota: Perkin-Elmer Corporation, 1979.

In: Lithium Batteries: Research, Technology... ISBN: 978-1-60741-722-4
Editors: Greger R. Dahlin, et al.

Chapter 8

ADVANCED LITHIUM-ION BATTERIES FOR PLUG-IN HYBRID-ELECTRIC VEHICLES

Paul Nelson[a] and Khalil Amine[b]
[a]Argonne National Laboratory, 9700 S. Cass Ave, Argonne, IL 60439, USA, 630-252-4503.
[b]Aymeric Rousseau, Argonne National Laboratory
Hiroyuki Yomoto, EnerDel Corp., Argonne, IL 60439, USA.

ABSTRACT

In this study, electric-drive vehicles with series powertrains were configured to utilize a lithium- ion battery of very high power and achieve sport-sedan performance and excellent fuel economy. The battery electrode materials are LiMn2O4 and Li4Ti5O12, which provide a cell area-specific impedance of about 40% of that of the commonly available lithium-ion batteries. Data provided by EnerDel Corp. for this system demonstrate this low impedance and also a long cycle life at 55°C. The batteries for these vehicles were designed to deliver 100 kW of power at 90% open- circuit voltage to provide high battery efficiency (97-98%) during vehicle operation. This results in battery heating of only 1.6°C per hour of travel on the urban dynamometer driving schedule (UDDS) cycle, which essentially eliminates the need for battery cooling. Three vehicles were designed, each with series powertrains and simulation test weights between 1575 and 1633 kg: a hybrid electric vehicle (HEV) with a

45-kg battery, a plug-in HEV with a 10-mile electric range (PHEV10) with a 60-kg battery, and a PHEV20 with a 100-kg battery. Vehicle simulation tests on the Argonne National Laboratory's simulation software, the Powertrain System Analysis Toolkit (PSAT), which was developed with MATLAB/Simulink, showed that these vehicles could accelerate to 60 mph in 6.2 to 6.3 seconds and achieve fuel economies of 50 to 54 mpg on the UDDS and highway fuel economy test (HWFET) cycles. This type of vehicle shows promise of having a moderate cost if it is mass produced, because there is no transmission, the engine and generator may be less expensive since they are designed to operate at only one speed, and the battery electrode materials are inexpensive.

Keywords: vehicle simulation, lithium-ion batteries, series-engine hybrid

1. Introduction

Lithium-ion batteries show promise for powering hybrid electric vehicles (HEV) and plug-in hybrid vehicles (PHEVs), but the batteries under development differ widely in their capabilities. Also, a variety of vehicle types are under consideration and the requirements for their batteries vary considerably: some demand high energy per unit volume and weight, and others place greater emphasis on high power. For these vehicle applications, the batteries are required to have safe and consistent performance throughout a life of about 15 years and be available in mass production at a moderate price.

2. Status of Advanced Battery Development

Table 1 presents the characteristics of several lithium-ion batteries in various stages of development. These batteries promise markedly different levels of performance for the various criteria for which batteries are evaluated for the HEV and PHEV applications.

The LiNi0.8Co0.15Al0.05O2 (NCA)-graphite system has good power and energy characteristics because of its high voltage, good electrode specific capacities and good area-specific impedance (AS I) [1, 2]. Projections show it would have a moderate cost in production and good life if the state of charge (SOC) is maintained between 90% and 30%. Work remains to be done to increase the useful fraction of the (SOC) range and achieve excellent battery life (15 years). Also, at present the NCA electrode has a tendency to release significant amounts of oxygen during thermal runaway, resulting in oxidation of the electrolyte [3,4]. The graphite electrode adds chemically bound energy to such a catastrophic incident.

The LiFePO4 (LFP) electrode is more stable and does not generate oxygen during heating and, thus, appears to be safer at this time than the NCA electrode. Otherwise, the performances of the first two systems in Table 1 are expected to be similar, but the LFP-graphite system shows promise of a slightly lower raw material cost. However, developing a low-cost process for preparing nano-LiFePO4 material, which is required for good power, requires additional effort.

The third system in Table 1, MS-TiO, has electrodes with low capacity, and the couple has lower voltage than those of the first two systems. The lithium-spinel positive electrode does not form a good couple with a graphite negative electrode because manganese dissolves in the electrolyte and poisons the graphite electrode [5-8]. Against a lithium-titanate electrode, however, it forms a very stable couple, albeit with a low voltage. The titanate electrode has a voltage that is 1.5 V higher than that of lithium, whereas a good graphite electrode is only about 0.1 V higher. The combination of low voltage and low specific capacities for both electrodes in the MS-TiO system results in lower specific energy for the battery than for most lithium-ion systems. This is somewhat mitigated by the very stable performance and long cycle life for 100% discharges, as discussed below, which permit operating MS-TiO batteries over the SOC range of 100% to 10% for the PHEV application. Another favorable characteristic of the MS-TiO system is very low ASI, which results in very high power. The safety characteristics appear to be excellent; it is very tolerant of excessive voltage upon charging, with a much reduced likelihood of lithium deposition, and the stored chemical energy in the system is very low when compared with that of systems with graphite electrodes. As will be shown by the data that follow, the cycle life for the MS-TiO system is excellent. Cost projections are only tentative, but the MS-TiO system appears to have inherent advantages over the other systems in that its electrode materials have low cost and are plentiful.

The fourth system in Table 1, MNS-TiO, is similar to the MS-TiO system, but with a manganese- nickel spinel positive electrode that operates at a very high voltage versus lithium (4.8 V at full charge) and with improved capacity relative to manganese spinel. At the present time the ASI is higher than for the MN-TiO system, but it is believed that this can be improved sufficiently to achieve the required power (100 kW for the PHEVs in this study) for a 40-mile range PHEV.

Table 1. Selected Lithium-Ion Battery Systems for Plug-in Hybrid Electric Vehicles.

System	NCA-Graphite	LFP-Graphite	MS-TiO	MNS-TiO	MN-Graphite
Electrodes Positive Negative	LiNi0.8Co0.15Al0.05O2 Graphite	LiFePO4 Graphite	LiMn2O4 Li4Ti5O12	LiMn1.5Ni0.5O4 Li4Ti5O12	Li1.2Mn0.6Ni0.2O2 Graphite
Capacity, mAh/g Positive Negative	155 290	162 290	100 170	130 170	275 290
Voltage, 50% SOC	3.6	3.35	2.52	3.14	3.9
ASI for 10-s Pulse, ohm-cm^2	25	25	9.2	100	25
Safety	Fair	Good	Excellent	Excellent	Excellent
Life Potential	Good	Good	Excellent	Unknown	Unknown
Cost	Moderate	Moderate	Low	Moderate	Moderate
Status	Pilot Scale	Pilot Scale	Develop.	Research	Research

The highest capacity positive electrode in Table 1 is in the MN-graphite system developed at Argonne [9,10]. This would result in the lowest battery weight for a 40-mile PHEV for the batteries reviewed in Table 1. This system requires more development work, but it illustrates the improvements in battery performance that may come in the future.

3. Spinel-Titanate Battery Performance Modeling

3.1 Approach

Despite its low capacity and low voltage, we have studied the MS-TiO system to determine if a battery-vehicle combination could be found that exploits the very high power of the MS-TiO system. A type of vehicle that may be particularly enhanced by very high battery power is one with a series-connected powertrain with sport-sedan performance. For such a vehicle to achieve high fuel economy, the engine should operate close to its peak efficiency, which requires that the battery have high power to accept charging at a high rate. Therefore, we decided to design a battery that could discharge at the 100-kW rate for a 10-s burst at 90% open-circuit voltage (OCV) so that the overall battery efficiency would exceed 97% for most vehicle-driving cycles. The high battery power would also make possible higher vehicle performance than is usually expected of HEVs and more like that of a sport-sedan.

Through a collaboration between Argonne and EnerDel, experimental data became available that establish the low area-specific impedance of the MS-TiO system and the promising long cycle life for deep discharges, which justify the assumption that the battery can be operated between 100% and 10% SOC. Two types of modeling were required to characterize the battery for vehicle simulation studies: (1) design modeling to determine the battery weight, volume and electrical performance and (2) impedance modeling. The experimental data and modeling are discussed below.

3.2 Experimental Data

Tests with a 1 .8-Ah MS-TiO cell demonstrated outstanding power; 97% of the capacity measured at the 1C discharge rate was delivered at the 50C rate (Fig. 1) [11]. These results were correlated to obtain the impedance equations required for the vehicle simulation tests.

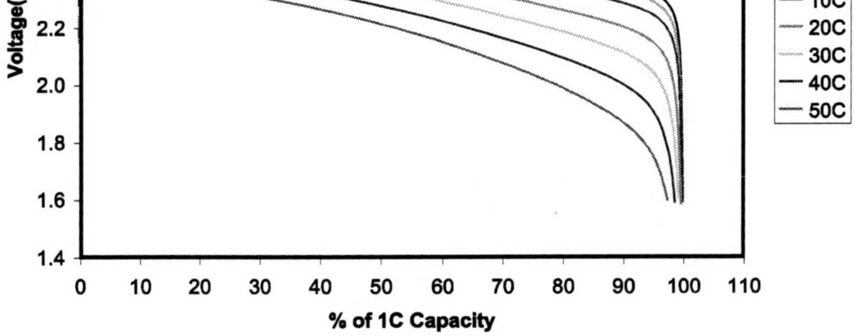

Figure 1. Lithium-Manganese Spinel/Lithium-Titanate 1 .8-Ah Cell Charged at 1C Rate and Discharged at Varying Rates at 30°C [11]

The capacity stability was demonstrated in tests in which the entire cell capacity was discharged and charged at the 5C rate at an elevated temperature of 55oC to accelerate degradation. After 2,300 cycles there was little indication of capacity loss (Fig. 2) [11]. Pulse power characterization tests were carried out at 30oC after 1,000 and 2,000 cycles and demonstrated little loss of power with cycling, and incidentally, restored the full initial capacity. The promising results obtained in these aggressive tests at high temperature indicate that MS-TiO batteries may be able to achieve the 5,000 cycles required for the PHEV application.

3.3 Battery Design Modeling

We have developed a method, based on Excel spreadsheets, for designing cells and batteries that has been applied for several battery systems. In recent years, the method has been used primarily for designing lithium-ion batteries for HEVs and PHEVs [12,13]. One form of input for this method is test results from measurements of capacity and ASI on small cells with areas of only a few square centimeters. It is also possible to accept data from larger cells by accounting for the resistance of the current collection system in the tested cells. The method calculates the volumes and weights of all of the cell and battery components and the electrical performance of the battery. By this method, three batteries were designed for a series-connected vehicle from the data in Table 1 for the MS-TiO system and from other proprietary input. The results are shown in Tables 2 and 3.

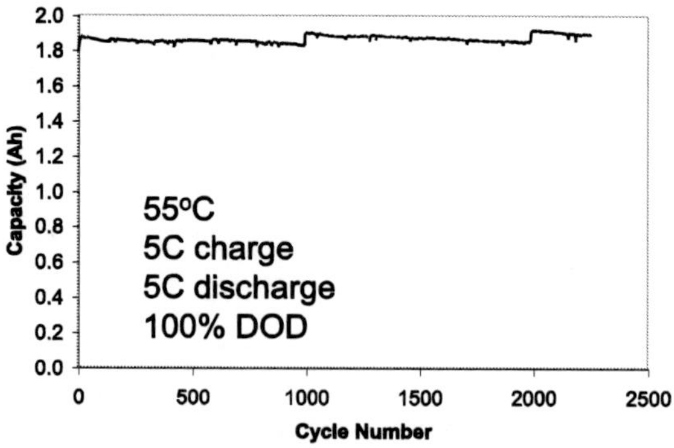

Figure 2. Deep Discharging of Lithium-Manganese Spinel/Lithium-Titanate Cell to Demonstrate Long Cycle Life [11]

Table 2. Cell Parameters for Lithium-Manganese Spinel/Lithium-Titanate Batteries for HEVs and PHEVs.

Cell Parameters	HEV	10-Mile* PHEV	20-Mile* PHEV
Cell Capacity (1/C rate), Ah	10.0	16.6	33.3
Positive First Charge Loading Density, mAh/cm^2	0.54	0.88	1.79
Negative-to-Positive 1st Charge Capacity Ratio	1.0	1.0	1.0
Maximum Voltage on Charging, V	2.7	2.7	2.7
Average Voltage on Discharge, V	2.51	2.51	2.51
Positive Electrode			
Active Material	Li1 .06Mn1 .94O4	Li1 .06Mn1 .94O4	Li1 .06Mn1 .94O4
Thickness of Coating (each side), μm	25	40	82
Negative Electrode Material			
Active material	Li4Ti5O12	Li4Ti5O12	Li4Ti5O12
Thickness of Coating (each side), μm	21	34	70
Total Cell Area, cm^2	20,500	20,500	20,500
Cell Dimensions, mm			
Height	189	219	219
Width	104	116	187
Thickness	12.2	12.4	12.5
Cell Weight, g	471	648	1102
Power, W	1251	1251	1251
Cell Specific Power, kW/kg	2.66	1.93	1.14
Cell Specific Energy (1/C rate), Wh/kg	53	64	76

*Based upon energy usage of 300 Wh/mile.

Table 3. Battery Parameters for Lithium-Manganese Spinel/Lithium-Titanate Batteries for HEVs and PHEVs.

Battery Parameters	HEV	10-Mile* PHEV	20-Mile* PHEV
Number of Cells in Battery	80	80	80
Number of Modules (10 cells each)	8	8	8
Energy Storage (1-h rate), kWh	2.0	3.3	6.7
Useable Energy			
HEV, 60% to 35% SOC	0.50		
PHEV, 100% to 10% SOC		3.0	6.0
Discharge Power (10 s), kW	100	100	100
Discharge Voltage af Full Power (50% SOC), V	181	181	181
% of Open Circuit Voltage	90	90	90
Power Density, kW/L	3.59	2.81	1.81
Current on Discharge, A			

Table 3 – (Continued)

Battery Parameters	HEV	10-Mile* PHEV	20-Mile* PHEV
At Rated Power (50% SOC, 90% OCV)	552.5	552.5	552.5
Maximum Allowed (30 s)	560.0	560.0	560.0
Maximum Regeneration Power, kW			
Short-Term (2-s regen braking)	100	100	100
Long-Term (up to 60 s)	70	70	70
Maximum Charge Voltage, V	216	216	216
Insulated Battery Wall Thickness, mm	7	7	8
Battery Dimensions, mm			
Length	852	973	973
Width	266	270	274
Height	123	135	207
Volume, L	28	36	55
Weight, kg	45	60	100
Total Weight of Cells, % of Battery Weight	84	86	88
Cooling Fluid (exterior of modules only)	Air	Air	Air

*Based upon energy usage of approximately 300 Wh/mile.

3.4 Impedance Modeling

On the basis of the data shown in Figures 1 and 2, the impedance of the experimental cell was modeled to fit Equation (1) in Figure 3 [14].

Equation (1) $1000*(OCV-V_L)/I_L = R = Ro+Rp_1*Ip_1/I_L+Rp_2*Ip_2/I_L$

Where, OCV = open circuit voltage, V

VL = cell voltage, V

R = total cell impedance, milliohms

Ro = cell internal ohmic resistance, milliohms

Rp1 = first internal polarization resistance, milliohms

Rp2 = second internal polarization resistance, milliohms IL = cell load current, A

Ip1 = current through first polarization resistance, A

Ip2 = current through second polarization resistance, A

The values for Ip1 and Ip2 are derived by integration of the differential equation:

Equation (2) dlp/dt = (IL-Ip)/ô

The results were adjusted to simulate a 1 -Ah cell to facilitate use in calculating battery impedance for any desired capacity and number of cells and are shown in Table 4, in which the values of Tau1 and Tau2 are time constants expressed in seconds for the two polarization resistances in the model.

The impedance parameters for the 1-Ah Cell of Table 4 were applied for the 100-kW batteries of Table 3 with the result illustrated in Table 5.

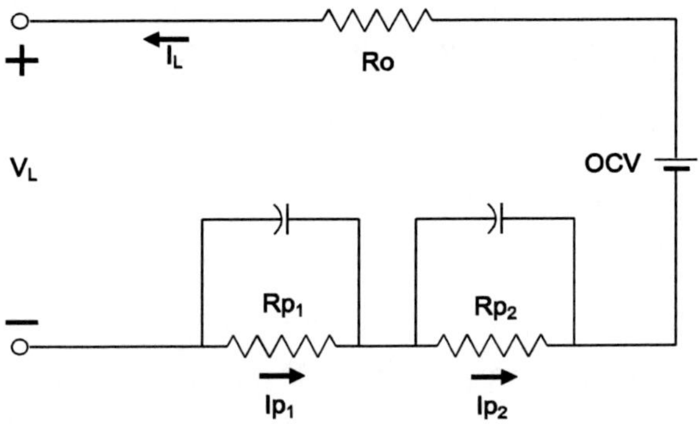

Figure 3. Impedance Model for Lithium-Manganese Spinel/Lithium-Titanate Batteries [14]

Table 4. Parameters for Calculating Impedance of a 1-Ah Lithium-Manganese Spinel/Lithium-Titanate Cell.

DOD, %	OCV	Ro	Rp1	Rp2	Tau1	Tau2
0	2.661	0.00320	0.00220	0.00100	10	270
10	2.621	0.00320	0.00220	0.00120	10	270
20	2.593	0.00320	0.00209	0.00130	10	270
30	2.569	0.00320	0.00220	0.00130	10	270
40	2.543	0.00320	0.00230	0.00140	10	270
50	2.514	0.00320	0.00266	0.00140	10	270
60	2.483	0.00320	0.00313	0.00132	10	270
70	2.446	0.00320	0.00355	0.00108	10	270
80	2.408	0.00320	0.00420	0.00100	10	270
90	2.368	0.00320	0.00500	0.00100	10	270
95	2.336	0.00380	0.00600	0.00100	10	270
100	1.6	0.00440	0.00700	0.00100	10	270

Table 5. Impedance, Voltage, and Current for 10-second Power Burst for 1 00-kW Lithium-Manganese Spinel/Lithium-Titanate Batteries

10-s Burst Discharge at 100 kW					10-s Burst Power at 560 A, kW
SOC, %	R-10s	V	%OCV	A	
100	0.0342	195.4	91.8	511.8	108.5
90	0.0343	191.8	91.5	521.3	106.7
80	0.0338	189.6	91.4	527.3	105.6
70	0.0343	187.2	91.1	534.2	104.3
60	0.0348	184.6	90.7	541.7	103.0
50	0.0364	181.0	90.0	552.5	101.2
40	0.0386	176.8	89.0	565.6	99.1
30	0.0405	172.2	88.0	580.8	96.9
20	0.0435	166.5	86.5	600.4	94.2
10	0.0472	160.0	84.4	625.2	91.3
5	0.0562	149.2	79.9	670.1	87.0
0	0.0652				

4. Vehicle Simulation for High-Power Batteries

4.1 Approach

The Powertrain System Analysis Tool (PSAT) [15, 16], developed with MATLAB/Simulink, is a vehicle-modeling package used to simulate performance and fuel economy. It allows one to realistically estimate the wheel torque needed to achieve a desired speed by sending commands to different components, such as throttle position for the engine, displacement for the clutch, gear number for the transmission, or mechanical braking for the wheels. In this way, we can model a driver who follows a predefined speed cycle. Moreover, as components in PSAT react to commands realistically, we can employ advanced component models, take into account transient effects (e.g., engine starting, clutch engagement/ disengagement, or shifting), and develop realistic control strategies. Finally, by using test data measured at Argonne's Advanced Powertrain Research Facility (APRF), PSAT has been shown to predict the fuel economy of several hybrid vehicles within 5% on the combined cycle. PSAT is the primary vehicle simulation package used to support the U.S. Department of Energy's (DOE's) FreedomCAR research and development activities.

4.2 Vehicle Characteristics

Several vehicles were sized for different specifications based on the same vehicle attributes:

- HEV
- PHEV with 10 miles All Electric Range (AER)
- PHEV with 20 miles All Electric Range (AER)

The main component masses are shown in Table 6 and Table 7 lists the main characteristics of the simulated midsize car.

As shown in Figure 4, the configuration selected is a series engine hybrid, very similar to the one used in the GM Volt [17].

Five driving cycles are considered in the study to evaluate the impact of advanced lithium-ion batteries on fuel economy: UDDS (urban dynamometer driving schedule), HWFET (highway fuel economy test), LA92 (1992 test data from Los Angeles), NEDC (new European driving cycle) and Ford ATDS. The main characteristics of each cycle are summarized in Table 8.

Table 6. Mass of Vehicle Components (kg)

Component	HEV	PHEV10	PHEV20
Engine Mass	120	120	120
Generator Mass	86	86	87
Motor Mass	144	144	146
Battery Mass	45	60	100
Vehicle Mass	1575	1590	1633

Table 7. Vehicle Main Specifications

Component	Specifications
Engine	2004 US Prius
Electric machine	Ballard IPT - Induction
Single Gear Ratio	2
Final Drive Ratio	3.8
Frontal Area	2.1 m2
Drag Coefficient	0.25
Rolling Resist.	0.007 (plus speed related term)

Wheel radius	0.317 m

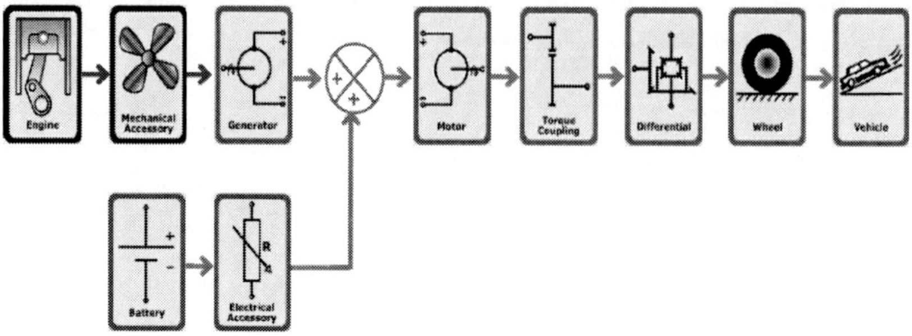

Figure 4. Series Engine Configuration.

Table 8. Drive Cycle Characteristics.

	Unit	UDDS	HWFET	LA92	ATDS	NEDC
Duration	S	1372	764	1435	1799	1180
Distance	Km/mi	11.92/7.45	16.38/10.24	15.7/9.81	25.2/15.75	10.9/6.84
Average Speed	mph	19.5	48.26	24.6	31.5	20.86
Average Accel	m/s2	0.5	0.19	0.67	0.55	0.59
Average Decel	m/s2	-0.57	-0.22	-0.75	-0.55	-0.78
Number stops		17	1	16	18	13
Percent Stops		18.92	0.65	16.3	20.73	24.9

Note that all the simulations performed in PSAT represent hot conditions.

4.3 Component Sizing Algorithm

The components of the different vehicles were sized to meet the same vehicle performances:

- 0-60 mph in less than 7sec
- Gradeability of 6% at 65 mph

To quickly size the component models of the powertrain, an automated sizing process was developed. A flow chart illustrating the sizing process logic is shown in Fig. 5. While the engine power is the only variable for conventional vehicles, HEVs have two variables: engine power and electric power. In that case, the engine is sized to meet the gradeability requirements while the battery is sized to meet the performance requirements. In the study, we also insure that the vehicle can capture the entire energy from regenerative braking during decelerations on the UDDS.

Similar to the HEV configuration, the engine and generator powers are sized to meet the gradeability requirements. In addition to HEVs, the battery power has to be sized to follow the UDDS driving cycle while in all-electrical mode. Finally, the battery energy is sized to achieve the required AER of the vehicle. The AER is defined as the distance the vehicle can travel on the UDDS without starting the engine. Note that a separate control algorithm is used to simulate the AER. This algorithm forces the engine to remain off throughout the cycle, regardless of the torque request from the driver.

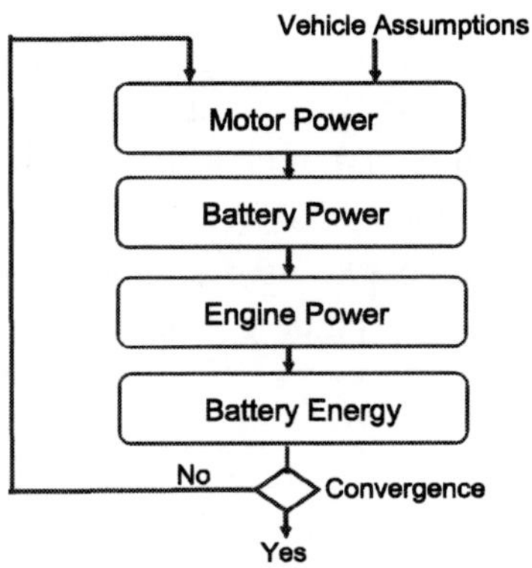

Figure 5. PHEV Component Sizing Process.

The main component characteristics resulting from the sizing algorithm are described in Table 3.

Table 9. Component Sizing Results.

		HEY	PHEY10	PHEY20
Engine Power	kW	100	100	102
Generator Power	kW	95	95	96
Motor Power	kW	130	130	132
Battery Power	kW	100	100	100
Vehicle Mass	kg	1575	1590	1633
Accel. Time 0-60 mph	s	6.2	6.2	6.3

4.4 Control Strategy Philosophy

The control strategy of the PHEVs can be separated into two distinct modes, as shown in Figure 6:

- Charge-Depleting (CD) Mode: Vehicle operation on the electric drive, engine subsystem or both with a net decrease in battery SOC.
- Charge-Sustaining (CS): Vehicle operation on the electric drive, engine
- subsystem or both with a 'constant' battery state-of-charge (i.e., within a
- narrow range), which is similar to that in current production HEVs.

During a simulation, the engine is turned on when the battery SOC is low or the power requested at the wheel cannot be provided by the battery alone. Turning the engine on expends fuel but conserves battery energy, so that more miles can be traveled before the battery reaches its discharged state. When the engine is ON, it is operated close to its best efficiency curve. As a result, the battery is being charged by the engine during low power requests, leading to lower electrical consumption.

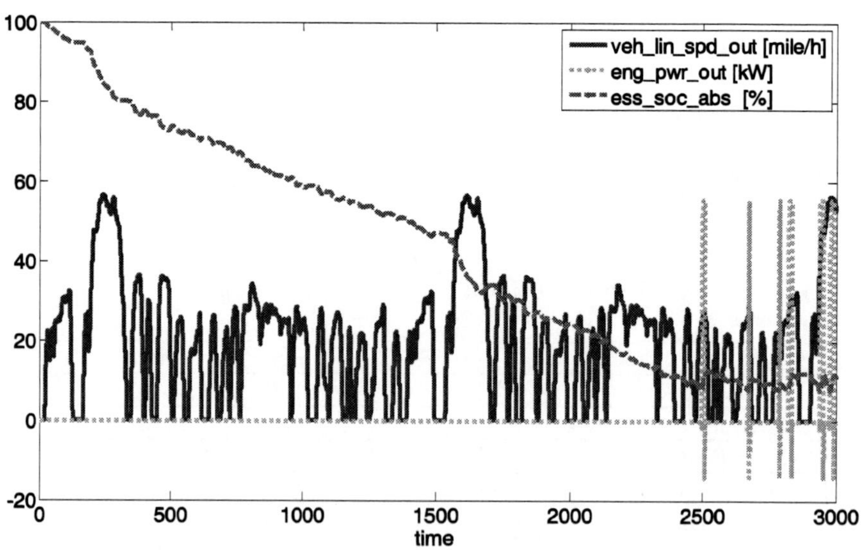

Figure 6. Control Strategy SOC Behavior on.

The initial SOC of the battery, which is also the battery's maximum charge, is 100%, and the final SOC of the battery, which is also the battery's minimum charge, is 10%. For the CD mode, the engine logic was written in StateFlow and used several conditions, such as battery SOC, motor power limits, and vehicle speed, to determine when the engine should turn on and the output torque of the engine. The logic of the CS mode was similar to that of current HEVs.

4.5 Fuel Economy Results

As previously mentioned, several driving cycles have been considered to evaluate the benefits of the advanced lithium-ion batteries on PHEVs. Table 4 summarizes the electrical consumption on each PHEV vehicle on the first cycle of each drive cycle. These results highlight the differences between the different drive cycles. As expected, the standardized drive cycles (UDDS, HWFET and NEDC) require a lower electrical consumption than the cycles that are more "real-world". The ATDS is the most aggressive drive cycle.

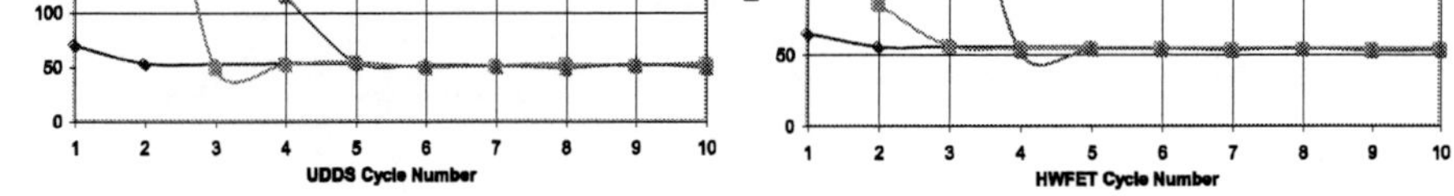

Figure 7. Fuel Economy on UDDS and HWFET.

Table 10. PHEV Electrical Information.

		UDDS	HWFET	NEDC	LA92	ATDS
10 miles AER	Elec Cons. First Cycle (Wh/mile)	224.6	204.3	234.1	282.6	190.4(1)
	All Electric Range (miles)	13.8	14.3	12.8	10.3	9.5
20 miles AER	Elec Cons. First Cycle (Wh/mile)	257.9	209.9	241.6	297.9	300.8
	All Electric Range (miles)	26.6	28.6	26.5	20.4	19.9

(1) Engine started during the first cycle

Because the primary goal of PHEVs is to maximize the fuel displacement, the following analysis focuses on fuel consumption. Figure 7 shows the evolution of the fuel economy when each cycle is repeated 10 times.

The benefit of high-power batteries is noticeable on the more aggressive driving cycles (Fig. 8). When an engine start would have been necessary for low power batteries, the initial distance can be performed in EV mode without any help from the engine. Note, however, that previous studies [18] have demonstrated the need to know the trip distance to properly minimize fuel consumption. However, higher battery power allows additional flexibility in deciding when to start the engine.

Table 5 shows the charge sustaining fuel economies of the different vehicles. Due to increased vehicle mass, the fuel economy decreases slightly with an increase in All Electric Range.

Figure 9 shows the evolution of the electrical consumption for the UDDS and ATDS drive cycles. The impact of the cycle aggressiveness can be seen by the slope of the electrical consumption. In the case of the ATDS, the slope is much stiffer than for the UDDS.

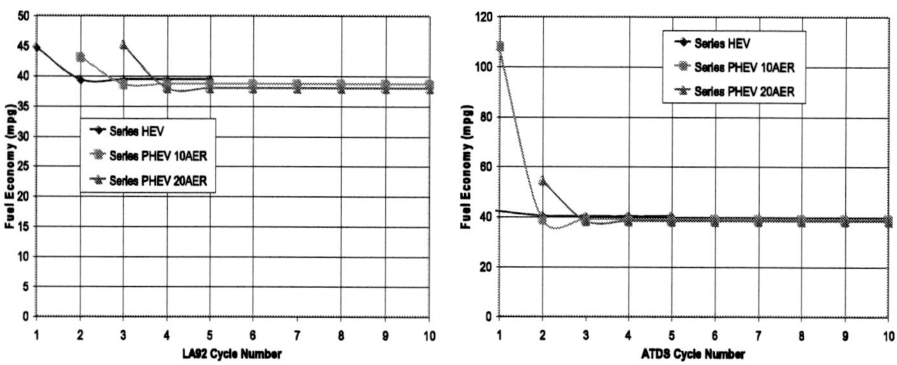

Figure 8. Fuel Economy Evolution on LA92 and ATDS

Table 11. Charge Sustaining Fuel Economy (mpg)

	UDDS	HWFET	NEDC	LA92	ATDS
HEY	51.9	54.4	52.3	39.3	40.0
PHEY 10	51	53.3	51.5	38.6	38.8
PHEY 20	49.6	52	50.5	37.9	38

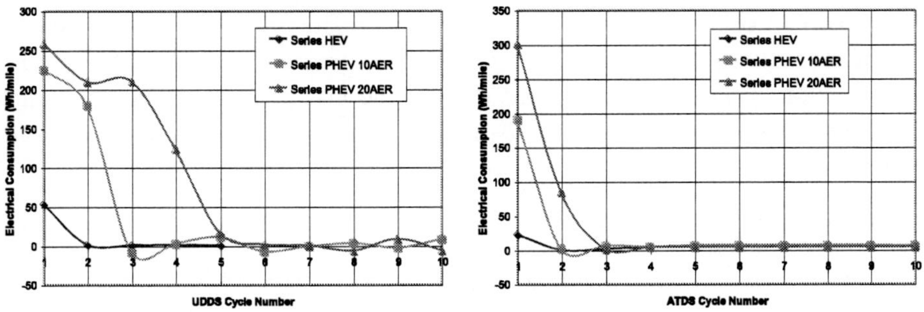

Figure 9. Electrical Consumption Evolution on UDDS and ATDS

Table 11. Component Average Efficiencies (%) on UDDS

Component	HEY	PHEY10	PHEY20
Engine	36.9	37.2	37.2
Generator	91.9	91.9	91.9
Motor	80.4	80.4	80.4
Battery	98.4	97.5	97.4
Gear	97.5	97.5	97.5

The efficiencies of the vehicle components are very high as illustrated in Table 11 for theUDDS cycle. Improvement in the fuel economy for these vehicles could be achieved by increasing the motor efficiency. An additional motor of low power (30-50 kW) could be provided to be used under light loads under which it could operate at higher efficiency than the high-power motor (130 kW) in the evaluated designs.

The high battery efficiency, results in very little battery heating. One hour of travel on the UDDS cycle would heat up the PHEY10 battery by only1.6 oC under adiabatic conditions.

5. Conclusions

High vehicle performance, of the type expected from sport sedans, and high fuel economy can be achieved at the same time by a vehicle having a series powertrain and a high-power manganese spinel/lithium titanate battery. Further improvement in fuel economy might result from improving the motor efficiency. This battery can provide high power at such high battery efficiency that battery cooling is virtually unnecessary. This type of vehicle shows promise of having a moderate cost if it is mass produced because there is no transmission, the engine and generator may be less expensive since they are designed to operate at only one speed and power, and the battery electrode materials are inexpensive.

6. ACKNOWLEDGMENTS

This work was supported by DOE's FreedomCAR and Vehicle Technology Office under the direction of Tien Duong and David Howell of that program. The submitted manuscript has been created by the UChicago Argonne, LLC, Operator of Argonne National Laboratory ("Argonne"). Arrgonne, a U.S. Department of Energy Office of Science laboratory, is operated under Contract No. DE-AC02-06CH1 1357. The U.S. Government retains for itself, and others acting on its behalf, a paid-up nonexclusive, irrevocable worldwide license in said article to reproduce, prepare derivative works, distribute copies to the public, and perform publicly and display publicly, by or on behalf of the Government.

7. REFERENCES

[1] Amine, K; Chen, CH; Liu, J; Hammond, M; Jansen, A; Dees, D; Bloom, I; Vissers, D; Henriksen, G. *Journal of Power Sources.*, 2001, 97-8, 684-687.

[2] Abraham, DP; Liu, J; Chen, CH; Amine, K. *Journal of Power Sources.*, 2003, 119, 511-516 Sp. Iss. SI.

[3] Belharouak I; Lu, WQ; Vissers, D. et al. *Electrochemistry Communications.*, 2006, 8(2), 329-335.

[4] Doughty, DH; Roth, EP; Crafts, CC; Nagasubramanian, G; Henriksen, G; Amine, K. *Journal of Power Sources.*, 2005, 146, 116-120.

[5] Amine, K; Liu, J; Belharouak, I. *Electrochemistry Communications.*, 2005 7, 669-673.

[6] Amine, K; Liu, J; Belharouak, I; Kang, SH; Bloom, I; Vissers, D. Henriksen, *Journal of Power Sources.*, 2005, 146, 111-115.

[7] Amine, K; Liu, J; Kang, S. et al., *Journal of Power Sources.*, 2004, 129, 1, 14-19.

[8] Chen, Z; Amine, K. *Journal of the Electrochemical Society.*, 2006, 1 53(2), A3 1 6-A320.

[9] Kang, SH; Sun, YK; Amine, K. *Electrochemical and Solid State Letters.,* 2003, 6(9), A1 83-A1 86.
[10] Kang, SH; Amine, K. *Journal of Power Sources.*, 2005, 146, 654-657.
[11] Amine, K; Belharoauk, I; Liu, J; Tan, T; Yumoto, H; Ota, N. Abstract No: 2F09, 48th Battery Symposium in Japan.
[12] Nelson, P; Dees, D; Amine, K; Henriksen, G. *J. Power Sources.,*2002, 110(2), 349.
[13] Nelson, P; Bloom, I; Amine, K; Henriksen, G. *J. Power Sources.,* 2002, 110(2), 437.
[14] Nelson, P; Liu, J; Amine, K; Henriksen, G. ECS Proc. Vol. (F1) *Power Sources Modeling.,* 2003.
[15] Argonne National Laboratory, PSAT (*Powertrain Systems Analysis Toolkit*), *http://www*. transportation.anl.gov/.
[16] Rousseau, A; Sharer, P; Besnier, F. "*Feasibility of Reusable Vehicle Modeling: Application to Hybrid Vehicles,*" SAE paper 2004-01-1618, SAE World Congress, Detroit, http://www.eere.energy.gov/vehiclesandfuels, March 2004. http://www.gm-volt.com/
[17] Karbowski, D; Rousseau, A; Pagerit, S; Sharer, P. "Plug-in Vehicle Control Strategy: From Global Optimization to Real Time Application", 22th International Electric Vehicle Symposium (EVS22), Yokohama, October 2006.

INDEX

E

F

G

H

I

J

K

L

M

N

O

P

R

S

T

V

W

X